Cambridge Elements

Elements in the Philosophy of Physics
edited by
James Owen Weatherall
University of California, Irvine

THE PHILOSOPHY OF SYMMETRY

Nicholas Joshua Yii Wye Teh
University of Notre Dame

CAMBRIDGE
UNIVERSITY PRESS

Shaftesbury Road, Cambridge CB2 8EA, United Kingdom

One Liberty Plaza, 20th Floor, New York, NY 10006, USA

477 Williamstown Road, Port Melbourne, VIC 3207, Australia

314–321, 3rd Floor, Plot 3, Splendor Forum, Jasola District Centre,
New Delhi – 110025, India

103 Penang Road, #05–06/07, Visioncrest Commercial, Singapore 238467

Cambridge University Press is part of Cambridge University Press & Assessment,
a department of the University of Cambridge.

We share the University's mission to contribute to society through the pursuit of
education, learning and research at the highest international levels of excellence.

www.cambridge.org
Information on this title: www.cambridge.org/9781009507301

DOI: 10.1017/9781009008600

First published 2024

A catalogue record for this publication is available from the British Library.

ISBN 978-1-009-50730-1 Hardback
ISBN 978-1-009-00504-3 Paperback
ISSN 2632-413X (online)
ISSN 2632-4121 (print)

Cambridge University Press & Assessment has no responsibility for the persistence
or accuracy of URLs for external or third-party internet websites referred to in this
publication and does not guarantee that any content on such websites is, or will
remain, accurate or appropriate.

The Philosophy of Symmetry

Elements in the Philosophy of Physics

DOI: 10.1017/9781009008600
First published online: May 2024

Nicholas Joshua Yii Wye Teh
University of Notre Dame

Author for correspondence: Nicholas Joshua Yii Wye Teh, nteh@nd.edu

Abstract: This Element is a concise, high-level introduction to the philosophy of physical symmetry. It begins with the notion of "physical representation" (the kind of empirical representation of nature that we effect in doing physics), and then lays out the historically and conceptually central case of physical symmetry that frequently falls under the rubric of "the Relativity Principle," or "Galileo's Ship." This material is then used as a point of departure to explore the key hermeneutic challenge concerning physical symmetry in the past century, namely understanding the physical significance of the notion of "local" gauge symmetry. The approach taken stresses both the continuity with historically important themes such as the Relativity Principle, as well as novel insights earned by working with contemporary representational media such as the covariant phase space formalism.

Keywords: symmetry, Noether's theorems, representation, art, physics

ISBNs: 9781009507301 (HB), 9781009005043 (PB), 9781009008600 (OC)
ISSNs: 2632-413X (online), 2632-4121 (print)

Contents

1 Introduction

The notion of "symmetry" has beguiled physicists and natural philosophers from Aristotle to Einstein, and within the contemporary practice of physics, its hold on the physical imagination has only continued to grow in its depth and potency. This means that "symmetry" is also a topic of primary concern for the philosopher of physics, who wishes to elucidate the role that fundamental physical concepts – such as symmetry – play within the empirical practice of physics.

Despite the myriad ways in which various *mathematical* symmetries find application in subfields as disparate as quantum field theory, classical mechanics, General Relativity (GR), fluid mechanics, quantum gravity, and condensed matter physics, there is nonetheless a conceptually central theme within the history of physics that concerns the relevance of these symmetries to the empirical modeling practice of physics – a theme that runs powerfully from Galileo through Newton and Huygens, and then through Einstein, and then onward to their intellectual descendants (and as we will see, the theme has an important precursor in the verse of the Song dynasty poet Chen Yuyi and his boat). It is fitting, then, that the philosophy of physics literature has tended to focus on this conceptually central case, which is typically discussed under the head of "Galileo's ship" or the "Relativity Principle" (RP), and on the possibility of extending it to novel scenarios, for example to the local gauge symmetries of electromagnetism or the diffeomorphism symmetries of Einstein's GR.

This Element is meant to be a concise introduction to the philosophy of this conceptually central understanding of symmetry. Sections 2 and 3 lay out the minimal background about "physical representation" (the kind of empirical representation of nature that *we effect in doing physics*) that I will need, and introduce the notion of "symmetry" – and our central theme of Yuyi's boat (or Galileo's ship) – from this perspective.

Sections 4 through 6 introduce the reader to the question of whether the local symmetries that are so prevalent in contemporary physical theories can be thought of as continuous with the tradition of representation marked by the theme of Yuyi's boat – that is to say, can they too be empirically significant in the paradigmatic way discussed in Section 3? Although this question has been widely debated by philosophers (see e.g. Greaves and Wallace (2014), Teh (2016), and references therein), my discussion contains two novelties. First, unlike most of the extant literature, I discuss this question from the perspective of the formal conception of symmetry introduced by Emmy Noether, which is very much the *de facto* conception of symmetry in contemporary physical practice – as we will see, this move not only makes the conceptual issues more relevant to practitioners but also pays dividends to our conceptual analysis. Second, I link the discussion of this question – its complications and well as its resolution – to Einstein's search for a substantive notion of general covariance, and in particular, to his response to Felix Klein in the famous Klein–Einstein dispute (philosophers have of course discussed the issue of "substantive general covariance" a great deal, but to the best of my knowledge, no one seems to have explicitly linked this discussion to the empirical significance of symmetry).

Finally, it would be a shame if a high-level introduction such as this one did not give the reader a sense of the frontiers of present research: this is left to the Epilogue, where I sketch how recent work on "edge modes" connects our central theme with topics such as spontaneous symmetry breaking and the equivalence principle.

The intended audience of the Element is philosophy and physics graduate students (or advanced undergraduates) who already have a working knowledge of physics/mathematics, and some exposure to philosophical argumentation. So, for instance, in Sections 5 and 6 of the Element, I will assume that the reader already has an understanding of tensors, the covariant derivative, differential forms, the exterior derivative, and the Hodge star operator (all fairly elementary topics that one might expect to encounter in a good undergraduate class on the geometry of physics). On the other hand, Sections 2 through 4 really assume very little of the reader by way of mathematical prerequisites. Sections 2 and 3 in particular should be accessible to anyone who has an interest in the topic and a little physical maturity (e.g. a sense of the kind of implicit "effective" or scale-dependent reasoning that comes into play whenever we do physics).

2 Physical Representation

Section 2.1 sets up the background view of "physical representation" – the tradition of representing (or modelling) physical phenomena in which physicists are engaged – that I will be assuming in my subsequent discussion of symmetry in physics. This background view is important to my discussion for at least two reasons. First, it motivates and frames the particular way in which (in Sections 3 and 4) I present the shift from representation that uses global symmetries (i.e. symmetries that do not depend on spacetime) to the search for an analogous representation that uses local symmetries (i.e. symmetries that are nonconstant functions of spacetime). Second, it offers one way of making intelligible the significance of a monumental development in physicists' understanding of symmetry: as I argue in Sections 4 and 5, it helps us understand how and why the powerful *mathematical* rearticulation of symmetry by Emmy Noether (1918) came to so profoundly shape the *physical* task of pursuing physical representation in terms of local symmetries, and – as I explain in Section 6 – to play a role in resolving the attendant hermeneutic difficulties.

Much of this Element's discussion of representational relevance of symmetries is inspired by the seminal work of Harvey Brown (H. Brown, 2005; H. R. Brown and Sypel, 1995) on physical symmetry, and especially his discussion of the RP. Those familiar with Brown's work will know that this is only one of the two central pillars on which it rests – the second is a view often referred to as "the dynamical approach to spacetime," also developed in (Brown, 2005). In Section 2.2, I provide a brief note explaining that – at least in my view – a core insight (albeit one that is less emphasized by subsequent commentators) of Brown's dynamical approach is very much kindred with my discussion of physical representation.

2.1 General Background

This Element is about *physical* symmetry: that is to say, the notion of symmetry that is implicated in the *empirical* practice of physics. Within this practice, we pursue an understanding of certain aspects of the world – an understanding that is correlative to prediction and confirmation – by *representing* these aspects in particular ways, and by employing and elaborating on (one might even say: by *performing*) these representations. Thus, while the theme of "representation" is not my main topic, it is an essential piece of background for appreciating the notion of physical symmetry that I am after. Without some grasp of what "physical representation" – namely, the kind of representation practiced by physicists – amounts to, it would be impossible to latch onto the distinctive subject of *physical* symmetry (as opposed to mathematical symmetry, or symmetry in some other representational practice such as painting or sculpture), and it would be impossible to formulate the distinctive philosophical questions that pertain to this subject.

I would thus like to begin by briefly sketching my views on physical representation, in anticipation of Section 3's discussion of how symmetry comes to play a role in physical representation.[1] Since my view is highly analogous to the view that Podro (1998) advanced in the case of depiction (especially for

---

[1] While some may disagree with me about the details of how physical representation works, these disagreements should not prevent them from appreciating much of what I have to say about physical symmetry in the later sections of this Element (in particular, many of my morals will have correlates that can be articulated within their own preferred account of representation). At any rate, my aims are modest: I wish to sketch just enough background about physical representation to make it clear to the reader how symmetry – in the context of physical representation – is actually experienced, taken up, and deployed by practicing physicists.

Figure 1 *Portrait Study of Sir Thomas More*. Black and colored chalks on unprimed paper, 38×25.8 cm, Royal Collection, Windsor

drawings, paintings, and relief sculptures), it will help to introduce the view by first considering representation rather more generally, and getting a feel for how "idealization" – a theme of great interest in contemporary philosophy of science – plays out in an especially accessible case such as a painting.

With that in mind, let us begin with the idea of "representation" or "imitation" or "imaging" more generally. The *locus classicus* for a philosophical discussion of representation is of course Book X of Plato's Republic: why, Socrates archly inquires of the slow-witted Thrasymachus, would anyone want to view a representation of a subject if they were in a position to view the actual subject? The implicit assumption here, which Thrasymachus does not think to question, is that representations are mere (albeit unachievable) attempts to provide a literal or "isomorphic" copy of the subject, and are thus always inferior to directly experiencing the subject.

The reason Socrates' question is liable to strike us as absurd is that it flies in the face of so much of what we know about how representations are designed to work, and how they are appropriately received. Take the case of a representation such as the following Holbein sketch of St Thomas More at Windsor Castle (Figure 1).

Here the stroke is loose, open, and relaxed; the blending is soft, and the tint is light; furthermore, these various elements come together so that – as

Martz (1990) notes – we see in the sketch "the face of a man unguarded, open, vulnerable, seeking, devotional in its mood" (it is instructive to compare this with Holbein's more famous portrait of More at the Frick, which conveys perfect composure and an iron-clad determination).

Pace Socrates, what I would like to draw your attention to is the implausibility of construing Holbein's sketch as aiming at literal verisimilitude – as approximating the visual experience of gazing at the actual More. To understand this point, consider first the sketch's medium (chalk and brown wash on paper) and its procedures (the loose stroke that deliberately omits detail), and then consider how these are put to representational use. The sketch does not try to disguise its medium and its procedures as it would if it were aimed at simulating the experience of seeing More; on the contrary, when we contemplate the sketch, it is an intended part of our experience that we are aware of the medium – and the quality of its strokes – *as* a medium; we are conscious that an aspect of what confronts us is a surface marked with very particular qualities of stroke.

An important moral here is that while the medium and its procedures (as such) can be said to have their own "meanings" (the impulse of this stroke, the torsion of that spiral, the *impasto* of the oil paint), the proper or "literal" domain of these meanings is the marked surface. On the other hand, in a representation these meanings are *nonliterally* or metaphorically transferred to a new domain, namely the subject (St Thomas More, in this case). Thus, the sketch can be said to "idealize" or "distort," because the qualities of the medium are in the literal sense inapplicable to the representation's subject, and must thus suffer loss in relation to more literally veridical modes of confronting its subject (actually seeing More, for instance).

In the typical case of a sketch, we are meant to understand that the meanings of the medium do not literally apply to the subject; furthermore, as Vasari (1900) first observed in his discussion of artistic "design" or *di segno*, the representational power of the sketch is rooted in this understanding, as well as our grasp of the literal/proper meanings of the medium: for instance, it is through our *following the formulation of Holbein's stroke* as loose, open, and relaxed that we come to experience the sketch's distinctive power to disclose something about its subject, namely More's openness and vulnerability. The point is that through this particular metaphorical transference of meaning, we come to perceive or *think* the unity that is the representation – we do not experience "the medium's procedures" and "the face of St Thomas More" separately (nor indeed this in addition to some long chain of conditions relating the two in order to simulate a unity), but rather a fusion of the two, whose unity governs our responses. As Scruton writes (of portraits in general), "...I respond to the flowing lines and flesh-tints with emotions and expectations that derive from my experience of faces, and to the face with emotions and expectations that arise from my interest in colour, harmony, and expressive line." (p. 87 of Scruton (1997)).

It is this metaphorical transfer, I would suggest, that gives rise to much of the distinctively powerful content of representations – a content that is not propositional, and *cannot be had* in this particular way apart from the embodiment of thought *in* the representation, and a content that is (among other things) received in the kind of thought that we call contemplation. In Podro's (1987) poetic rendering of this point, a representation "...directs itself to the mind of the perceiver, who sees the subject remade within it, sees a new world which exists only in the [representation] and can be seen only by the spectator who attends to the procedures of [the medium]".[2] And this, ultimately, is why we find it absurd of Socrates to suggest that a representation such as Holbein's sketch aims to copy the literal appearances: Socrates is denying the very feature – the intended idealization and metaphorical transfer effected by the distinctive procedures of the medium and our awareness of them – that gives rise to the power of these representations.

Of course, the case can be considerably more complicated than that of Holbein's portrait of More, as Podro (1998) discusses at length in his monograph *Depiction*. A particular choice of medium and its procedures (e.g. the technique of linear perspective) conditions the representational possibilities that are open to the artist, foreclosing some of these to a greater or lesser extent and at the same time opening up or suggesting new possibilities; indeed a certain drawing or painting procedure (think of linear perspective or pointillism) may take on a life of its own, lending a seemingly inescapable momentum to the act of representation and helping to extend our thought about the subject. And it may be that this conditioning itself complicates the act of representation and the search for a compelling representation of the subject (consider for instance the struggle to maintain realism of animation in the transition from the Florentine *trecento* to the more optically accurate paintings of the *quattrocento*). Nonetheless, some of the most remarkable representational successes in painting have been borne out of cases where these complications have been bent toward achieving a compelling representation of the subject – in Sections 4–6, we will see how a central thread in the development of physical symmetry can be understood in this way.

---

[2] NB: Podro's point is not that the subject is thereby fictionalized, but rather that the subject comes to be understood in a way that is unique to that representation, as opposed to what we can understand through confronting it "face to face," or in some other kind of representation.

There is of course much more that could be said about how drawings and paintings represent. But the example of Holbein's sketch will suffice to illustrate the general features of representation that will inform my approach to physical representation.

(i) A representation involves a medium and a subject.

(ii) A representation is an embodied, performative kind of thought – we use the medium and its procedures to think the subject *in* the representation, and in so doing, we reenact the pattern of attention of the representation's maker(s), itself embodied in that medium and its procedures.

(iii) The *di segno* thesis: in a representation, we come to understand the subject by following the procedures of the medium.[3] It is through this fruitful interplay between the represented subject and our awareness of the medium *qua* medium that a representation distinctively extends and mobilizes our thought about its subject: it absorbs the particular momentum of the medium and its procedures and uses these to reconstruct the subject as can only be done in the world of that representation.

Let us see how this conception plays out in the case of physical representation. First, let us fill in (i) by specifying the "subject" and "medium" of physical representation. I take it that the focal *subject* of a physical representation is the dynamics of an empirical scenario, namely the description of how certain empirical degrees of freedom evolve over time. These dynamical degrees of freedom could be particle positions and velocities, or field configurations, or even quantities in theories whose formal description does not involve dynamical equations of motion (such as thermodynamics).[4]

The particular forms that such an empirical scenario could take are various, but very roughly, all of them should have some conception of the target degrees of freedom whose evolution we would like to keep track of – call these the *subsystem* degrees of freedom – and some reference standard that makes the description and identification of the subsystem degrees of freedom an empirical matter – call the system that defines this reference standard the *environment* (note that on this conception, the specification of an environment is correlative to the subsystem that one wants to model). The environment system might consist of spatially localized and separated degrees of freedom from the target

---

[3] Reciprocally, we come to experience the medium and its procedures under the aspect of the subject.

[4] For instance, in the case of thermodynamics, our static equations are still being used to tell us something about empirical dynamics – what happens (at some scale) after we lift the wall that separates two chambers, or push the piston, etc.

subsystem (e.g. some material body far away from the subsystem) as is often assumed in textbook mechanics, and – for historical reasons – this is indeed the case that I will focus on in the rest of this Element. At the same time, it is important to note that my use of the term is in principle much more general: it can encompass nonspatially separated (from the subsystem) fields of various kinds and local inertial frames – anything that provides a reasonable reference standard in the context of some measurement setup will do, and the particular details of what an environment system looks like will be highly theory dependent.

Since these are empirical scenarios, the subsystem-environment configuration needs to form what Cartwright (1999) calls a "nomological machine": a "...fixed (enough) arrangement of components, or factors, with stable (enough) capacities that in the right sort of stable (enough) environment will, with repeated operation, give rise to the kind of behavior that we represent..." Furthermore, the subsystem degrees of freedom should be thought of as implicitly indexed by a certain scale (e.g. length or energy scale) in relation to the environment and our measurement scale – typically, the ratio of scales set by this relationship that marks out the physical regime of interest and makes some particular set of degrees of freedom an empirically sensible description (it also gives us a qualitative grip on what "enough" means in the quote from Cartwright). This is not to say, of course, that in latching on to the subject one needs to have highly theoretical thoughts about how to quantify different regimes and describe precisely how they are related (as we do, for instance, when using the set of techniques known as *effective field theory*), but anyone seriously involved in emperically effective representation nonetheless has a loose – perhaps quite informal – sense that there is a regime involved, and of some of the bounds of this regime. In order for these representations to be taken up in use, we assume that there is an observer with access to the kinds of scales that define the measurement relationship between the subsystem and the environment; in the context of our subsequent exploration of physical symmetry, we will see that it also makes sense to introduce a subsystem observer, whose measurements are confined to the characteristic scale of the subsystem.[5]

The subject of a physical representation tends to be something that is informally conveyed in physics classes – especially those that emphasize empirical

---

[5] In general, the question of whether an observer should be modelled as part of the system or environment – and the degree to which this should be left implicit – will vary from context to context. In the context of our subsequent discussion of symmetry, the important point about observers is that one observer – the subsystem observer – should only be able to probe scales within the subsystem (which would not include the subsystem's relationship to the environment), and another – external – observer should have access to the relationship between the subsystem and the environment.

know-how and the interface between theory and experiment – and the laboratory, whereas textbook presentations tend to emphasize the *medium* of a physical representation: the *mathematical* apparatus (differential equations, geometry, coordinate frames, functions, sections of bundles, etc.) that is used to model the subject. Nonetheless, it is important to recall the point that I made earlier about a representation being a unity: we experience a physical representation as a unity, that is not as the mathematics of the model and the empirical subject merely conjoined, but rather as inextricably fused – in the world of the model, we understand and manipulate various pure mathematical objects (geometries, symmetries, equations) in a way that is inflected by our understanding of experimental protocols and know-how, and we respond to and conceptualize empirical scenarios in a way that is driven by our interest in mathematical objects.[6] Of course, in the process of describing, analyzing, or teaching some part of our representational experience, it is good and natural to home in on particular aspects of it (the mathematical framework, or some aspect of the experimental configuration etc.) – similar to what one does in a close reading of literature – but this does not change the fact that one's experience is of the whole.

With this in mind, let us proceed to the version of point (ii) that applies to physical representation: a physical representation is a kind of performative, embodied thought, in which we use the mathematical medium and its procedures (geometry, partial differential equations, the symmetries of these objects, etc.) to embody a certain pattern of attention toward an empirical scenario – the kind of attention that is concerned with understanding how to model and predict and measure, and to extend that understanding to novel scenarios. This is especially obvious when we consider how we are to receive a "dynamical equation" in a physics textbook: the mathematics that we confront is really a prompt to perform a characteristic action of a physicist, namely to use that equation (and boundary conditions, analysis of the stability and regularity of solutions etc.) and know-how about the link between theory, and experiment to model the empirical subsystem degrees of freedom – this is what it is to think *in* the representation.[7]

---

[6] Readers who are familiar with the notion of "hylomorphism" in the writings of Aristotle will immediately see that my discussion resembles his discussion of "substance" in the Categories and the Metaphysics; however, the kind of unity that I am discussing – though real – is meant to have a lower ontological status than that of substance.

[7] One may well wonder if this very practical and concrete view of physical representation can accommodate putative physical models that are so abstract/schematic that they do not plausibly seem to have any empirical scenarios as part of their representational content. For instance, one might wonder if a mathematical geometry – take bare Minkowski spacetime, to fix ideas – with a smattering of physical interpretation counts as a physical representation in this sense.

The point is well-illustrated by considering even the simplest subject, such as the motion of some medium-sized dry good (the subsystem object) with respect to some reference object in an environment, where (let us assume) speeds are small relative to the speed of light, and the scale of the investigation is one for which the subsystem object is going to count as "free" or "isolated," and its extension matters little. This is a paradigmatic scenario for physical representation, and one in which context we introduce the formal mathematical medium of a function $x^i : \mathbb{R} \rightarrow \mathbb{R}^3$ and an ordinary differential equation $\ddot{x}^i = 0$ whose solutions are straight lines in $\mathbb{R}^3$.[8] What it means, however, for us to actually be able to receive this physical representation is to take it up in use as physics: to understand that $x^i(t)$ represents the position of the subsystem object at time t and that a solution to $\ddot{x}^i = 0$ represents an empirically possible trajectory given a certain timescale; to understand that this position should not be taken literally as a position of a point (it is not part of the content of the representation that "there is an object that has no extension...") but rather as modelling the insight that the object's extension does not matter in the regime of interest to us; to understand that the environment has been left implicit in the formal medium, and that modelling space as $\mathbb{R}^3$ is not to represent it as infinite (as a confused literalistic reading might suppose) but is instead a way of dramatizing the "isolated" property of the subsystem; and to understand much else in a similar vein, including the relationship between the mathematical medium and how we are to actually prepare and measure the position and velocity of the subsystem object, that we are not actually positing that quantities have real-number values, and so on.

Finally, point (iii) – the *di segno* thesis – takes us deeper into the relationship between mathematical medium and subject in a physical representation, and how this relationship works to extend our physical understanding of the subject, and even to reimagine and transform the subject quite radically within the world of physical representation. The idea here is that mathematics has its own rhythms, textures, and procedures (e.g. its own impetus to generalize its objects, such as moving from flat space to curved space) and that we come in part to perceive the subject – the empirical scenario – within the

---

To this, I would say that although the kinds of cases that I have been describing are *focal* for physical representation, there is also such as thing as *generic* representation – for instance, Xu Beihong's galloping horses are not representations of any particular horses, but rather function as abstract icons of a sort. I think there is also room within physical representation for such icons; nonetheless, our physical understanding of such representations ultimately turns on our ability to discern some kind of relationship between these and the focal cases (cf. also Cartwright's discussion of fables and models in Cartwright (1999)).

[8]   Here $\dot{x}$ denotes a derivative with respect to the domain of the function x, and so it is interpreted as a time-derivative in the context of mechanics, where the domain is the "time line."

representation through appreciating the medium and procedures of a physical representation *as* mathematical, and through the metaphorical transfer of these (mathematical) meanings to the subject, which yields the distinctive content of the representation.[9]

To give an example of which H. Brown (2005) makes much, and which will be a persistent theme in much of what follows, it is easy to consider the equation $\ddot{x}^i = 0$ and notice – at a mathematical level – that its form does not change under rotations, uniform boosts and translations in three-dimensional Euclidean space; and that these transformations take solutions to solutions. This is an example of mathematical symmetry (invariance of a mathematical object under mathematical transformations), but when it is taken up into the kind of understanding involved in physical representation, we come to perceive something physical in following the mathematical procedure of symmetry: the way in which various empirically possible subsystem trajectories are related to each other, and further – as we will discuss in Section 3 – the inability of a subsystem observer to discriminate between these trajectories.[10] In a manner analogous to Holbein sketch – where we come to perceive the openness of More's disposition by following the openness of Holbein's stroke – here we come to understand an empirical invariance (of what one can observe at a certain scale) by following the formulation of a mathematical invariance.

In the next section, I will say more along these lines about *the* historically and conceptually central case of symmetry in physics, namely the RP and how it exemplifies the notion of physical representation that we have just been discussing. On the other hand, recall the more subtle and interesting possibility that I discussed earlier in the case of painting, namely that introducing a novel procedure or modification of a representation's medium might create a fruitful complication for the work of resolving the representation's subject; of seeking and finding the subject again in the act of representing with novel technique. One might thus wonder: Are there examples of this kind of scenario playing

---

[9] At this juncture, it is worth pointing out a key way in which my account of physical representation differs from what some might consider a more "mainstream" account of scientific representation, such as the one offered by Callender and Cohen (2006), who hold that scientific representation is a derivative kind of representation that can be reduced to representation of a more primitive sort (e.g. mental representation) by an act of stipulation. On my account, the kinds of (physical) representations I have been discussing cannot be so reduced, because the particular choice of medium is constitutive – via metaphorical transfer – of the (in general nonpropositional) content of the representation.

[10] Actually, there is something even more subtle here: since we do not prepare and measure real number-valued initial and final data, but rather a range, we do not actually use a single mathematical solution to represent an empirical trajectory, but rather a neighborhood of such a solution. Thus, the relationship between nearby solutions is already in play when we are modelling a particular empirical trajectory.

out in the case of physical representation? In Sections 4–6, I will make the case that a pivotal episode in the development of physical symmetry (and its resolution) is best understood as a scenario of just this kind. The episode that I am thinking of is Einstein's attempt to introduce a novel procedure of the mathematical medium (local spacetime symmetry) and to seek and rework the subject of the RP within this new representation, as well as Emmy Noether's subsequent triumph: a supremely fluent rearticulation of the mathematical procedure that Einstein initiated.

To conclude this sketch of physical representation, let me say a few words about how the notions of "idealization" and "physical theories" fit into this picture. First, on idealization: the term "idealization" is used in many ways in the philosophy of science literature, but here is what I take to be a fairly neutral and uncontroversial description from Frigg and Hartmann (2020): "idealized models are models that involve a deliberate simplification or distortion of something complicated with the objective of making it more tractable or understandable." Notice that this broad description is not yet committed to any particular theory of what the content of an idealized representation is, and whether we should understand that content as making a false claim about the target of the representation.

From what I have said thus far, it should be clear that idealization in this broad sense is inherent to what I call a physical representation, because any choice of medium will introduce a divergence between the subject-as-represented and the subject itself. Nonetheless, on my account, this sense of idealization is not aptly fleshed out as a "representation as if," that is "representing the subject as if it were something it in fact is not" in virtue of idealizing. To use an example that I already raised earlier, it is not part of the representational content of a formal point particle model of a ball that the ball is in fact a point; on the other hand, it is part of the (perhaps formally implicit) representational content that the empirical regime we want to model – the *subject* of the representation – is one such that it makes sense to invoke the formal device of a point.

I should note that some writers, for example Potochnik (2020), do offer an account of idealization as "representation as if" and thus take the important phenomenon of idealization to count against the verisimilitude of the theory. While I agree with Potochnik about the importance and pervasiveness of idealization, on my view of what physical representation is, the problem is not verisimilitude but rather a *literalistic* conception of verisimilitude: one that fixates on the mathematical syntax of (the medium of) a physical representation and then takes that to naively disclose the content or ontological claims of a representation ("the represented object is a point..."). By contrast, in my view the focal content of a physical representation is nonliteral and one should adopt a conception of verisimilitude that aligns with that content.[11]

Second, on physical theories: I have said quite a bit about physical representation thus far, and not much about "physical theories" at all. Broadly speaking, I agree with Wallace (2021a) that as used in physics, the notion of a "theory" is "...ambiguous between various elements of a complex hierarchy"; furthermore, the notion tends to be used very formally (and without much concern for representation) in some quarters of the literature. For instance, on one level, one might define a theory as the collection of solutions to some equation of motion having fixed the boundary conditions (a formalistic approach to this definition is quite popular in some parts of the philosophical literature, where the "solutions" are referred to as 'models'). And at the next level up, one might take a theory to allow for different choices of boundary conditions (thus modelling different subsystem-environment relationships). These and more general uses all have their role to play in the practice of physics, provided that one does *not* proceed purely formally and always considers the empirical-representational role that the relevant mathematical devices are playing. For our purposes, it will suffice to think of "a theory" as a collection of physical representations in the above sense, joined by a particular subject matter (e.g. gravity) and accompanied by a keen "know-how" about what we can do with such representations and how they are related to each other.

In the subsequent sections, we will be plunging deep into what I earlier called "the margins of analysis" of a representation, where – in an analogous way to any serious analysis of a painting – we will inevitably find ourselves immersed in commentary on subtle and technical features of the mathematical medium, while trying not to lose our sense of the representation as a whole. When we are in the throes of such necessary – but potentially disruptive – analysis, I ask the reader to bear in mind that what ultimately interests us is physical representation in the sense that I have just laid out.

2.2 The Dynamical View

In the next section, we will be entering more deeply into a discussion of how the concept of symmetry enters into physical representation, and in particular, one that draws heavily on Harvey Brown's work on the (representational) interpretation of physical symmetry (H. R. Brown and Sypel, 1995). Although it is well-known that this is only one of the two key strands of Brown's work – the

---

[11] This is of course not to underplay the importance of mathematics in the representational practice of physics; on the contrary, it is to give it the room to do the work it in fact has to do, as I hope our case study in Sections 4–6 will make clear.

other being "the dynamical approach to spacetime" – the relationship between the two is rarely remarked on, even by Brown himself. In this brief interlude, I would like to point out that the understanding of physical representation that I have sketched above can be understood as capturing some of the core insights in Brown's "dynamical approach to spacetime," as advanced in H. Brown (2005) – indeed I take it to be a way of elaborating on Brown's view that is not only deeply faithful to the motivations behind the dynamical approach, but explains its continuity with Brown's views on physical symmetry and will thus provide important background for our discussion of the RP in Section 3. Nonetheless, I should acknowledge that my gloss on Brown runs contrary to the mainstream interpretation of the dynamical view by subsequent commentators such as Read (2018) (on the other hand, it is somewhat closely related to the interpretation of Brown given by Stevens (2020), albeit without the Humeanism).

The dynamical approach is most simply stated for the theory of Special Relativity (SR), so I will confine myself to this setting here. Read (2018), a prominent expositor of the dynamical approach, introduces it by first defining the view it is opposed to, namely what he calls the "geometrical approach": according to Read, the geometrical approach holds that "...the Minkowski metric field η_{ab} of SR is ontologically autonomous and primitive, and [...] constrains the possible form of dynamical equations for matter, such that metric symmetries coincide with dynamical symmetries." (Here, what Read means by "metric symmetries" is those mathematical symmetries that leave η_{ab} invariant, collectively known as the Poincare group.) By contrast, Read writes of the dynamical approach that on "...this view, the metric field η_{ab} is not ontologically autonomous and primitive; rather, it is a codification of the symmetry properties of the dynamical equations governing matter fields." (Here, the relevant frame-based special relativistic equations are left form-invariant under the action of the Poincare group, which is of course of a piece with the fact that η_{ab} is used to write them in a frame-invariant form.) According to the dynamical approach, "dynamics" (in a sense that we will soon explore) is prior to, and is codified by, our interpretation of the mathematical entity η_{ab}.

The point that I wish to highlight is that there are two ways to further elaborate on the dynamical view beyond the negative claim that the mathematical entity η_{ab} (the Minkowski metric) should not be taken to (quite literalistically, I assume) represent some ontologically autonomous entity. The first way is already suggested by Read when he writes that "One may, therefore, understand the dynamical approach to SR—and to theories with fixed metric structure more generally—as an ontological thesis; as a form of relationalism." Read is suggesting that the dynamical approach is itself making a positive ontological claim, where this might perhaps be further cashed out in terms of some ontological entities that are represented (literalistically, let us assume) by the dynamical equations and their symmetry properties, but not by the metric.

Nonetheless, there is a second way of understanding Brown's claim that in SR, the Minkowski metric η_{ab} is merely a codification of the dynamics, and that is to understand "dynamics" here in a representational mode: as the evolution of the subsystem degrees of freedom of an empirical scenario (which presumes the background context of an environment, a particular regime of interest involving certain length scales, time scales, and measurement accuracy, etc.). From this point of view, the metric η_{ab}, the equations of motion, the relevant boundary conditions, and the Poincare symmetries – various aspects of the mathematical medium of the representation – are *all* codifications of the dynamics of an empirical scenario. Furthermore, while this representational reading of Brown's dynamical approach is compatible with various ontologies, one thing that is clear on this reading is that the mathematical medium of a physical representation should not be literalistically interpreted as suggesting some particular ontology. Understood in this way, the dynamical approach is not a statement about ontology at all.[12]

Our treatment of symmetry will be very much of a piece with this understanding of the dynamical view: the mathematics of symmetry plays a key role in the (nonliteralistic) representation of an empirical scenario, and in this capacity, it should not be taken to suggest any particular ontology (although it might in fact be compatible with various ontologies).

2.3 Further Reading

The view of physical representation that I have sketched here has a close affinity with those of several other authors writing about "modeling and representation" in the general philosophy of science. For instance, it is very much the spirit of the chapter titled "Fables and models" in Cartwright (1999), and also related to certain themes in Chang (2022), Van Fraassen (2010), and Anscombe (1971). As a reference on the philosophy of scientific representation, I recommend Frigg and Nguyen (2020), and for a detailed study of how representation

---

[12] While one can of course find statements in Brown that support the former interpretation (i.e. one that makes a positive ontological claim by virtue of some literalistic understanding of representation), my preferred version is not without its interpretive merits: for instance, Brown holds that the kinematics and the dynamics of a theory are intertwined, and it is much easier to make sense of this position on my view than on the former view. Furthermore, while the former view tends to treat SR and GR as theories of the universe *tout court*, this is difficult to square with Brown's emphasis on subsystems and his view of RP; on my interpretation, Brown's treatment of the RP ends up being fully compatible with the dynamical approach to spacetime.

works in painting (by which my sketch has been greatly influenced), I refer the reader to Podro (1998). Finally, for a discussion of the concept of "space-time" that is very sensitive to its use in physical representation, I recommend H. Brown (2005).

3 Physical Symmetry

In the last section, I sketched a view on which the distinctive power of physical representation lies in a judicious metaphorical transfer of the meanings and rhythms of mathematics (the representation's medium) to an empirical scenario (the subject), which I further spell out in terms of a subsystem-environment configuration. In this section, I wish to offer an efficient introduction to how a central tradition of thinking about physical symmetries powerfully recruits mathematical symmetries in this way, in order to dramatize the observational scales in play in an empirical scenario, as well as their relationship to the dynamics that is being modelled.

Section 3.1 introduces an early version of the RP that was noticed by the Song dynasty poet Chen Yuyi, and proceeds to review H. R. Brown and Sypel's account of how standard physical representations of the RP work. I then turn to a more recent discussion of the RP (Greaves and Wallace, 2014; Teh, 2016) in terms of "empirical significance" – this discussion has not only been influential in articulating the RP, but also in setting out a key hermeneutic task for the philosophy of symmetry, namely extending the representational meaning of the RP to a more sophisticated notion of mathematical symmetry, often called "local symmetry." In Section 3.2 – which is propaedeutic to Section 4 – I explain how the rest of the Element will take up this hermeneutic task, and in particular why we will pursue a somewhat different strategy than Greaves and Wallace (2014) and my own previous work in Teh (2016). Finally, Section 3.3 briefly mentions some other work that has been done on the philosophy of symmetry, but to a rather different – non-representational, in my sense – end. I explain why much of this work is orthogonal to my inquiry, and highlight a marginal area of conceptual overlap.

3.1 Yuyi's Boat

I should like to initiate our discussion of physical symmetry proper by attending to – what is to the best of my knowledge – the earliest clear articulation of how the intuitive notion of physical symmetry figures in our grasp of the subject of a physical representation. In 1118, the Song dynasty poet Chen Yuyi penned the following stanza about sailing on the Grand Canal to Xiangyi, a suburb of Bianjing (here I use the translation offered by McCraw (1986)):

Flying flowers on both banks shine the boat red;
A hundred leagues along the elm dike; half a day of wind.
I lie and watch a sky full of motionless clouds;
Unaware that the clouds and I are both going east.

Putting aside the intricacies of poesy, the remarkable thing to notice is that Yuyi has here articulated a core aspect of the "ship thought experiment" discussed by Galilei (1967), and that subsequent generations of commentators would discuss under the rubric of the RP.[13] Put in a more familiar idiom, Yuyi has noticed that in this scenario, there is an observational scale (the "subsystem scale") according to which a traveler on the boat cannot tell the difference between being in a particular constant velocity state and an alternate scenario in which we "boost" or transform the boat into a distinct constant velocity state; and on the other hand, the poet also implicitly recognizes that there is an environment (the bank of the Grand Canal) equipped with an observational scale that is sensitive to the velocity of the boat relative to the bank, and according to which an external observer can discriminate between the boat being stationary, and the boat moving East with some nonzero velocity. In other words, from the perspective of the subsystem (and its observational scale), there is a subsystem symmetry – namely an invariance of what the "internal" or subsystem observer can detect under constant velocity boosts of his boat – but from the perspective of the total subsystem-environment configuration, there is a sense in which a symmetry fails to obtain; in a locution that I will only return to in the Epilogue, one might even say that there is a sense in which (from the total system perspective) a symmetry is "broken."

Evidently, the symmetry (and the lack thereof) that Yuyi has in mind here is a feature of the dynamics of the empirical scenario: of the composite subsystem–environment degrees of freedom. And this is true regardless of whether – as in Yuyi's case – one has a relatively rough and untheorized sense of this empirical dynamics, or whether, as in the cases that we are about to consider, one is viewing the dynamics under the more theorized aspect of "the equations of motion." It is no exaggeration to say that much of the history of physics has been an attempt to articulate and rework this understanding of symmetry, which is now referred to as the RP.

---

[13] I note that I am not suggesting that everything in Galileo's ship thought experiment is already present in Yuyi's observation; in particular, Yuyi's observation does not spell out the full range of conditions under which it is not possible for the subsystem observer (with his characteristic observational scale) to tell the difference between different motions.

How is the RP incarnated in what I earlier called a physical representation (see the discussion of Section 2)? In particular, how can some particular mathematical procedure be recruited so that in following its formulation *as* mathematics – and in understanding that it does not literally apply to the empirical subject – we come to reconstruct RP within the world of the representation, as a physical thought about a subsystem-environment configuration? I submit that we can find an answer to these questions in the work of H. R. Brown and Sypel (1995) on the philosophy of symmetry (a line of inquiry that has been carried out somewhat independently of Brown's work on the dynamical approach to spacetime, but is in fact closely related to it on my representational understanding of the dynamical approach, cf. Section 2.2).

In their rich and learned investigation, H. R. Brown and Sypel (1995) trace the history of attempts to articulate the RP from Galileo onwards, through Newton, Huygens, and Einstein. For instance, they note that in Corollary V of the Principia, Newton asserts a version of the RP: "The motions of bodies included in a given space [i.e. a subsystem, in our interpretation] are the same among themselves, whether that space is at rest, or moves uniformly forwards in a right line without any circular motion." The context here is that Newton was trying to recover Yuyi's observational subsystem symmetry from within the ambit of his chosen mathematical medium for representing the dynamics of certain empirical subsystem degrees of freedom, namely the equation of motion $\ddot{x}^i = 0$.[14] There is a remarkable continuity between Newton's statement and what we find Einstein writing almost two centuries later in the context of SR, viz. "The [equations of motion] by which the states of physical systems undergo changes are independent of whether these changes of states are referred to one or the other of two coordinate systems moving relatively to each other in uniform translational motion." (Einstein, 1905).

Of course, there are differences of detail here. For Einstein, the mathematical symmetry that expresses the subsystem observer's observational invariance is the Poincare group (the group of symmetries that leaves the Minkowski metric η invariant), whereas in Newtonian gravity the mathematical symmetry that expresses the subsystem observer's observational invariance is the Galilean group (the group of symmetries that leaves invariant the degenerate Galilean metric and a compatible flat connection).[15] But what Newton and Einstein's representational strategies have in common is much more important

---

[14] For a discussion of this attempted derivation and Newton's non-sequitur, see Section 3.2 of H. Brown (2005).

[15] Actually, it is somewhat anachronistic to lump together Newton or Einstein's statement of the RP with some particular choice of mathematical symmetry group. Their statements were more general than this, and Einstein in fact uses his version of the RP (along with the constancy of

than their differences – they both take a certain procedure of mathematical symmetry (their medium) and use it to dramatize a particular dynamical feature of an empirical scenario, namely the observational invariance of the subsystem observer and the corresponding observational variance of the external observer.

To give an explicit and very elementary example of the *mathematics* of symmetry that provides a medium for understanding RP (in the Einsteinian case, let us say), consider a coordinate-free (abstract index) version of the equation of motion for a free scalar field on a Minkowski background:

$$\eta^{ab}\nabla_a\nabla_b\phi = 0. \tag{1}$$

When we choose to write this equation in inertial coordinates (in which ∇ becomes an ordinary partial derivative), we see that although a general Poincare transformation will take us to a different inertial coordinate frame, the *form* of the equation is left invariant – a mathematical feature that is referred to as the "invariance" of the equation under the symmetry.

At this point, it is instructive to note that one *could* give a literalistic reading of the form-invariance of this equation under Poincare transformations: what we perceive in this representation is a lonely scalar field in the universe (represented by spatially infinite Minkowski space), and the different inertial coordinate frames are just further choices of representational convention on which no physical significance hangs; thus, no physical significance can be attributed to the form-invariance of an equation under transformations from one inertial frame to another. In consequence, either this form-invariance cannot be the expression of a physically contentful idea like RP, or – if it is – then RP cannot have any physical content. But this would be obtuse – as obtuse as looking at Holbein's sketch of More in the previous section and interpreting it as aspiring to photorealism (and judging it as such).

By contrast, H. R. Brown and Sypel (1995) offer a more perceptive account of how we come to recognize the RP by following the mathematical form-invariance of the scalar field equation (in an inertial frame). They emphasize two key points:

(1) By their lights, when Einstein writes that the special relativistic equations of motion (as expressed in an inertial frame) are form-invariant under the Poincare transformations (taking from one inertial frame to another), this

---

the speed of light) to *derive* the structure of his relativity group, as emphasized in (H. Brown, 2005; H. R. Brown and Sypel, 1995).

mathematical statement is really meant to dramatize the observational-invariance of the outcomes of subsystem experiments set up with the same initial and boundary conditions.[16]

In the language of the previous section, we can say that Einstein has created a physical representation in which we are intended to grasp the observational invariance of Yuyi's empirical scenario (adapted to SR) through following the mathematical formulation of "form-invariance of the SR equations of motion under Poincare symmetries," or alternatively, "the Poincare-invariance of the Minkowski metric used to formulate the SR equations of motion." Implicit in this representational understanding is the thought that "inertial frames" here are not mere coordinate systems, but also a stand-in for the subsystem observer – these inertial frames represent the subsystem observer's state of motion relative to the environment frame.

(2) They also stress that the formulation of Einstein's RP – and those of his seventeenth-century progenitors – assumes that the physical symmetry transformations (i.e. changes in the inertial state of motion) are performed on an *isolated* subsystem and are thus testable.

In other words, Brown and Sypel are highlighting the fact that it is not just any subsystem-environment that one is considering in a Yuyi's boat scenario, but rather one in which – due to the isolation of the subsystem – it makes sense to perform a subsystem symmetry that does not disturb the environment, and thus to compare the relational difference in the total (composite subsystem and environment) state before and after the subsystem symmetry is effected.

I will call the interpretation of RP given in (1) and (2) *Representational RP* because of its emphasis on how the mathematical symmetries (of equations of motion, in this case) need to be understood not literalistically, but as the medium of what I have called a physical representation: when we receive this representation appropriately, we use its mathematical procedure to reenact a certain pattern of attention – the one spelled out in (1) and (2) – toward an empirical subject. Such an understanding also makes it manifest why the RP is a *physical* principle and not a metaphysical or an a priori one.[17]

---

[16] Of course, there is no guarantee that any *formal* symmetry of a differential equation used to model some bit of physics will admit of such an empirically relevant representational interpretation. See Section 5 of Wallace (2019) for further discussion of this point, especially with respect to the Lenz-Runge symmetry.

[17] Of course, there is room for even more subtle interpretive questions here, such as whether a *gestalt* between the "mere coordinate change" *interpretation of (1) and its interpretation* as embodying RP might serve to further deepen and extend our imaginative thought about the RP, in much the same way that Rembrandt's etching of Jan Cornelis Sylvius – in which the

I now turn to a more recent discussion of Representational RP by Greaves and Wallace (2014), in which H. R. Brown and Sypel's analysis is re-articulated in greater generality, so as to highlight the generic conditions under which the mathematical symmetries of a physical theory can be used to represent "empirical significance" in the manner of Yuyi's boat. For instance, in the case of RP that we have just been discussing above, we focused on rigid spacetime symmetries, that is symmetries of a fixed background (non-dynamical) metric which are *rigid* in the sense that they are not functions of space and time. Greaves and Wallace's analysis clarifies the conditions under which any rigid symmetry (which need not be a spacetime symmetry) of a theory's Lagrangian or equations of motion can be used to exhibit empirical significance in the manner of RP. For the moment, we will follow Greaves and Wallace in considering symmetries of a theory's equations of motion, but in the next three sections, we will switch over completely to the more typical contemporary approach (within physics) of considering symmetries of Lagrangians – either way, the important point to take note of for now is that such symmetries send solutions (of the equations of motion) to solutions.

Let us now see how Greaves and Wallace's schema for Yuyi's boat-type empirical significance goes, to a first approximation. First, they introduce S, the space of subsystem solutions to the theory's subsystem equations of motion (equipped with "isolated" boundary conditions), and $\mathcal{E}$, the space of environment solutions to the theory's equations of motion, as well as the space of solutions $\mathcal{T}$ of the total (composite subsystem and environment) system; the rigid symmetries of S, $\mathcal{E}$, and $\mathcal{T}$ take solutions to solutions in each of these spaces, respectively.[18] Greaves and Wallace then tell us that a subsystem symmetry (that is, a symmetry of S) of an isolated subsystem has (Yuyi's boat-type) empirical significance just in case (i) it preserves the "isolated" boundary conditions (since these are a defining condition of that subsystem); and (ii) the solutions of $\mathcal{T}$ prior to, and after, performing the subsystem transformation are *not* related by a symmetry of $\mathcal{T}$.

Although this sounds like a mouthful, it is simply capturing Yuyi's observation – and Representational RP – in more formal and schematic terms, in which equations of motion and their solutions are used to represent the dynamics of an empirical scenario from the get-go.[19] To link Greaves and Wallace's

---

preacher's hand extrudes from the picture plane – uses the disruption of the boundary between literal and imaginative perception to further its representational ends.

[18] I refer the reader to Greaves and Wallace (2014) for a discussion of the compatibility conditions that have to be satisfied in setting up these spaces.

[19] Actually, starting with the equation of motion or Lagrangian is optional, as I will explain in the Epilogue: one could just as well describe a very coarse-grained description of the

schema back to point (1) of what I call the Representational RP in Section 3.1, we should note that implicit in their notion of a subsystem symmetry that preserve the boundary conditions – their condition (i) – is a "subsystem observer scale" with respect to which the subsystem observer cannot distinguish between subsystem experiments (set up with the same initial and boundary conditions) run in the pre- and post-transformed state. This can also be dramatized by saying that with respect to the measurement scale of the subsystem observer, the subsystem symmetry seems like a universe symmetry, which transforms everything in the same way and thus lead to any physical differences. And to explicitly link Greaves and Wallace's schema to point (2) of the Representational RP, their condition (ii) is simply a formal way of encoding the relational difference in the total system state before and after subsystem symmetry is performed.

It may help to give a concrete example at this point: consider a Newtonian particle (represented by a uniform velocity solution) deep inside a subsystem and far from the spatial boundary that divides the subsystem from the environment, in which there is, let us assume, some other reference particle. We can imagine boosting the subsystem particle to a different uniform velocity solution (relative to the reference particle) without changing anything significant (relative to the scales of interest) about the Newtonian gravitational potential at the spatial boundary; since we would have changed the relative velocity (between the subsystem and reference particles) by performing the subsystem boost, there is no total system symmetry (i.e. boost of the composite subsystem-environment system) that connects the relative velocity pre-subsystem-boost to the relative velocity post-subsystem-boost.[20] Thus, this lack of a total system symmetry is a manifestation of the relational differences that can be detected by an external observer.

Now, it is fairly uncontroversial that the rigid symmetry cases of Yuyi's boat have "empirical significance" along the lines of what Greaves and Wallace (2014) say. However, they are further concerned with two kinds of (seemingly unrelated) novel cases in which one intuitively wants to say that symmetries have some kind of empirical significance, but where it is not clear that the standard understanding of Representational RP, or Yuyi's boat-type empirical significance, straightforwardly applies.

---

empirical dynamics by means of symmetry – and without introducing equations of motion or Lagrangian – and use one's knowledge of a symmetry-breaking pattern to derive the effective Lagrangian (and thus equations of motion) for the subsystem. This technique – known as the "coset construction" – has recently been used to great effect by the condensed matter community (Delacrétaz et al., 2014).

[20] Note that, based on the composite system dynamics, a total system boost must preserve the relative velocities and so we can deduce that no such boost exists in this scenario.

The first case – and the important one for this Element – is one in which the symmetries (of the equations of motion or Lagrangian) are no longer rigid, but are now allowed to be "local," in the sense that they are nontrivial functions of space and time. To say this case is an important one in contemporary physics would be an understatement: it is this case that Einstein had to confront head-on in developing the theory of GR, and it is this case that obtains in electromagnetism, the Standard Model, and more generally, in various gauge theories that are used in both high energy and condensed matter physics. Greaves and Wallace (2014) claim that their Yuyi's boat-type schema can indeed be extended to apply to this case, provided that one appropriately specifies asymptotic boundary conditions that are preserved by the local subsystem symmetries (see Teh (2016); Wallace (2021b) for a further discussion of this extension, and the relevant boundary conditions).

The second case has a much older pedigree; it is already mentioned by Newton as Corollary VI of his Principia:

> If bodies are moved in any way among themselves, and are urged by equal accelerative forces along parallel lines, they will all continue to move among themselves in the same way as if they were not acted on by those forces. (1687b, p. 20.)

The idea is that here again there is observational invariance for the internal/subsystem observer, but not just with respect to uniform velocity boosts; in this novel empirical scenario, the subsystem observer is unable to discriminate between different (spatially) uniform *accelerations* of the subsystem with respect to an environment. Furthermore, such a transformation is not *obviously* a symmetry of Newtonian gravitation, at least when the theory is naively written.[21] Greaves and Wallace say that this case indeed has empirical significance, but of a type that is distinct from Yuyi's boat-type scenarios: they take its empirical significance to consist in the fact that the subsystem symmetry transformation (the uniform acceleration) does not preserve the subsystem boundary condition and is accompanied by a corresponding environment transformation such that the composite (subsystem-environment) state post-transformation belongs in $\mathcal{T}$ and is physically distinct from the original composite state.

I have discussed an interpretation of this claim in Ramírez and Teh (2020) and a further refinement has also been given by Wallace (2021b), but space

---

[21] In fact, these transformations are symmetries of both the Newtonian gravitational equations of motion and Lagrangian if the gravitational potential is understood to transform in the appropriate manner. I will comment more on this in the Epilogue.

constraints force me to put the Corollary VI-type case aside in the rest of this Element, at least until the Epilogue. In what follows, we will enter more deeply into the central question underlying the first case, namely the question of whether local symmetries can be used to extend our conception of Representational RP – sufficient unto the day is this representational complication introduced by the (mathematical) medium of local symmetry!

3.2 Our Task

I agree with Greaves and Wallace that understanding the empirical significance of *local* symmetry (e.g. internal gauge symmetry or diffeomorphism symmetry) is central to any philosophical attempt to understand physical symmetry as it is used in the representational practices of contemporary physics. However, instead of focusing on their abstract schema, I wish to adopt a somewhat different tack.[22]

There are three related reasons for this. First, focusing on this sort of abstract schema tends to obscure several important elements of how the mathematics of symmetry is bent to representational ends in the practice of contemporary physics: it abstracts away from the important notion of Lagrangian symmetries and its associated Noetherian machinery of currents and charges, which is common coin amongst most contemporary physicists.

Second, it elides an important way in which contemporary practice conceptualizes an isolated subsystem, namely as a subsystem that has certain kinds of conserved charges (by virtue of being isolated from external disturbances that would result in a "leakage" of such quantities into the environment). I should note that the term "charge" is being used here in a highly general way, that is not just to include electric charge, but also quantities such as linear and angular momentum, certain notions of mass, and so on. As a matter of fact, contemporary practice not only puts much theoretical weight on this concept of charge, but also put a huge operational emphasis – these subsystem charges play the role of observables that we can probe in various experiments.

Third, I want to focus on the practicing physicist's phenomenology of representation, and so instead of adopting an overly schematized – and thus potentially deracinated – way of articulating how local symmetries might produce complications for Representational RP, I would like to center our inquiry on an actual historical episode in which these complications became salient. Until the very early twentieth century, the tradition of physics had a relatively straightforward understanding of how the mathematics of rigid symmetry (symmetries

---

[22] See Teh (2016) and Ramírez and Teh (2020) for my discussion of their schema.

with constant parameters such as uniform boosts) serves to embody our under-standing of Yuyi's boat-type scenarios, or the Representational RP. But shortly thereafter, this Arcadian state of affairs was disrupted by the very internal momentum driving the use of this subtle medium – symmetry – in physical representation, and which led to the introduction of the *local* symmetries that we have just been discussing. In the episode that I wish to direct our atten-tion to, no less a figure than Einstein struggled with the task of understanding how – not just despite, but because of – the complications of local symmetries, he could re-construct within this novel medium the subject of Yuyi's ship in a manner at once familiar and strange.

It is perhaps a good time to flag for the reader that in most of the representa-tional tradition within which Einstein is working (and indeed in large swathes of present-day physics) there is no explicit modelling of the empirical scenario's environment in the mathematics of the representation, although the environ-ment is of course still represented, albeit by means of the boundary conditions of a subsystem (and the operational understanding of how those boundary con-ditions incorporate a standard of reference that comes from the environment). In fact, we already saw an example of this in H. R. Brown and Sypel's dis-cussion of how we should interpret Representational RP as embodied in the mathematical form-invariance of the scalar field equation on a Minkowski background – even though they have in mind an environment for the subsystem, this environment is only implicitly represented in the mathematical formal-ism; by contrast, Greaves and Wallace's discussion of Yuyi's boat makes this environment explicit, albeit in a highly schematized way.

In the next three sections, we will follow in the footsteps of Einstein, and thus we too will adopt the practice of incorporating our assumptions about the environment (and the subsystem-environment relationship) into the bound-ary conditions of the subsystem. In this sense, our representations will be less explicit than Greaves and Wallace's schema. Nonetheless, in a different sense, they will include much physically relevant detail that is absent from Greaves and Wallace's schema, namely detail about the structure of the space of solu-tions and the conserved quantities of a subsystem. These details are not only routinely appealed to by physicists (both formally and informally) but – as we will soon see – they are also essential for entering more deeply into the hermeneutics of local symmetry.

3.3 Some Orthogonal Themes

In the final part of this section, I should like to briefly explain why a certain subset of work within the philosophy of symmetry literature (which is typically presented as having "metaphysical" purport) is largely orthogonal to my topic of interest – and my line of inquiry – in this Element. On the one hand, it would be helpful to clarify for the student the reasons for this divergence; on the other hand, there is a sense in which – when suitably interpreted – some of these themes have analogs within our inquiry, and I will point to these connections.

In this Element, a physical "theory" tends to be associated (or even identified) with a class of mathematical models. These models are initially conceived of as "kinematically possible models," which are "tuples of specified geometrical objects" (cf. Martens and Read (2020)) that are unconstrained by equations of motion or a Lagrangian (meaning that we do not yet constrain them to be solutions to equations of motion, or critical points of the action defined by that Lagrangian). Given such a class of kinematically possible models, one can then further restrict to a set of mathematical objects called "dynamical models," which is the subspace of the kinematically possible models carved out by constraining the models to be solutions of the equations of motion, or to lie in the critical locus of an action.

I now turn to Martens and Read (2020) for a lucid presentation of a particular view about "symmetry" in terms of the above definitions, as well as several distinctions concerning symmetry that this literature finds to be important. First of all, Read and Martens are clear in their exposition that they initially treat the above definitions of models purely formally, namely as mathematics without any representational purport. And it is worth pointing out that this is already a different kind of treatment from what one is intuitively doing when one introduces a "kinematical space of fields" in a full-blooded physics setting: here, even though there is not yet any formal dynamics in the sense of equations of motion or an action, one is motivated to introduce such kinematical structures *in order to* represent the dynamics of an empirical scenario (even if, for now, that empirical scenario is a rather hypothetical or schematic one). In other words, the choice of kinematic structures is already one that is adapted to the project of representing a certain kind of dynamics.

Next, in order to avoid controversy surrounding what a symmetry is, Read and Martens introduce the minimal notion of a symmetry as transformations "...which (whether by definition or otherwise) are regarded as relating empirically equivalent models" (p. 7 of Martens and Read (2020)). At this point, representation is clearly on the scene, but if it is to be physical representation in the sense that I have established, then one needs to hear a lot more about the subject of the representation (in particular its subsystem-environment structure and the relevant scales in play) before one can arrive at a sensible judgment about whether two solutions are in fact "empirically equivalent." Since much of the literature in this vein seems content to ignore these details in its investigation of symmetry, its ends seem largely orthogonal to my own, a point that will further emerge as we now turn to the consideration of two pairs of approaches to symmetry that are discussed in this Element.

Again following the summary in Martens and Read (2020), I now go over the two pairs of approaches, followed by a brief discussion of their relationship to the topic of this Element, namely the philosophy of symmetry as it relates to physical representation.

First, the "interpretational" versus the "motivational" approach to symmetry. According to the interpretational approach, "two symmetry-related models of a theory typically may be regarded ab initio as representing the same possible world, even in the absence of a coherent explication of their common ontology" (p. 7 of Martens and Read (2020)). By contrast, according to the motivational approach, "the existence of symmetry-related models at most motivates us to provide an explication of the shared ontology of these symmetry-related models, but only once such an explication is provided is it legitimate to regard those models as representing the same possible world." (p. 8 of Martens and Read (2020)). If we take this description at face value – in which case empirical equivalence is a given prior to introducing the distinction and the notions of "possible world" and "sameness" are metaphysically loaded, then this dispute is one that is orthogonal to my representational ends.

Second, the so-called "sophistication" versus "reduction" approach to symmetry. This distinction is supposed to come into play when a physical theory has some mathematical structure – and typically symmetry structure – that one regards as "surplus"; an example that is often given (but which I disagree with, cf. Nguyen, Teh, and Wells (2020)) is the local $U(1)$ symmetry of electromagnetism. "Sophistication" is the position that, when confronted with this scenario, one should reformulate the theory's mathematics so that the models related by a symmetry are isomorphic; by contrast, "reduction" is the position that one should modify the theory's mathematics forming a naive quotient of the space of models by that symmetry.

Since this second pair of positions does bear some relationship to the project of physical representation as I understand it (and also some relationship to a long tradition of mathematical and physical theorizing prior to the introduction of these position), there is somewhat more to comment on concerning this distinction. First, there is the issue of how one should interpret the question of whether a certain structure is "surplus" or not. If the question is one of whether, for instance, the local $U(1)$ symmetry of electromagnetism plays a role in *physical* representation (as opposed to a role ascribed to it by *a priori* metaphysical or semantic considerations), then not only is the question relevant to the present inquiry, but it has also been a central topic of discussion and interpretation *within* physics in the last century. On this conception, whether a structure is to be classified as "surplus" is something that needs to be adjudicated with respect to the structure's representational use in the context of various empirical subsystem-environment configurations. And as Wallace (2019) has recently argued, from this practice-oriented perspective, reduction is only justified when one is trying to represent the internal degrees of freedom of a subsystem independently of any consideration of representing the coupling between this subsystem and other subsystems – in other words, it is justified (and useful) in a fairly limited range of contexts relative to our representational practice as a whole.[23] Furthermore, since Grothendieck's "Pursuing Stacks" and the introduction of the Batalin–Vilkovisky formalism roughly 40 years ago, mathematicians and mathematical physicists have on the whole shied away from taking the naive quotient of a space with symmetries acting on it, preferring instead to work with what is known as a "quotient stack" (or at the infinitesimal level, with an L^∞-algebra), which retains information about the different ways in which isomorphic objects are related (see e.g. Nguyen et al. (2020) for an elementary introduction to the philosophy of this topic).[24]

3.4 Further Reading

For a further discussion of symmetry in this vein, I refer the reader to H. R. Brown and Sypel (1995), Greaves and Wallace (2014), Teh (2016), Wallace (2019), Wallace (2021a), Wallace (2021b), and Ramírez and Teh (2020). The "subsystem-environment" view of symmetry has been challenged in Belot (2018) on the basis of a particular interpretation of the moduli space of t'Hooft-Polyakov monopoles; this interpretation has in turn been criticized in Wallace (2021b), but more work clearly remains to be done on both sides.

4 The Hermeneutics of Symmetry

In the last section, I reviewed how the mathematical device of rigid symmetries (symmetries that are not functions of spacetime) embodies the Representational RP, either by means of the form-invariance of the equations of motion, as in H. R. Brown and Sypel's discussion, or through rigid subsystem symmetries that send solutions to solutions, as in the Yuyi's boat-type schema discussed by Greaves and Wallace (2014). I also stressed how a central hermeneutic

---

[23] For further discussion of this point, see Gomes (2019, 2021), Nguyen et al. (2020), Rovelli (2014), and Teh (2015).

[24] At a more mundane level, one reason physicists have shied away from reduction is in order to be able to write local, Lorentz invariant action functionals. I thank an anonymous referee for this reminder.

problem for the philosophy of symmetry is to explain whether and how a convincing sense of the Representational RP can be conveyed by means of local symmetries, namely symmetries that are nontrivial functions of spacetime.

In the philosophy of physics literature, the discussion of whether local symmetries can embody the Representational RP has largely been conducted in terms of Greaves and Wallace's Yuyi's boat-type schema (on which, see Brading and Brown (2004), Greaves and Wallace (2014), Teh (2016), and the references therein). This is fine insofar as it goes, but, on the other hand, it is noteworthy that the philosophical discussion has largely been conducted in isolation from two themes that – both historically and conceptually – are completely central to the physical tradition's understanding of local symmetry. Thereby hangs a tale, and as we shall see, a further opportunity to shed light on Representational RP in the case of local symmetries.

The first theme is Einstein's struggle to interpret and understand "general covariance," which I will define in Section 4.2 (for a magisterial overview of "general covariance" and what it might mean, see Norton (1993)). This theme is straightforwardly related to RP: after all, general covariance is a local symmetry in the sense that I clarified earlier, and one of Einstein's chief goals in introducing general covariance is to extend the RP. In the rest of this section, I will leave aside Greaves and Wallace's Yuyi's boat-type schema for RP and focus on the complications arising from Einstein's own quest to extend RP; I will then reconnect this thread with Greaves and Wallace's schema in Section 5.2.

The second theme, which was historically bound up with the interpretation of general covariance – but less obviously related to the RP – is the significance of the Klein-Einstein dispute for the foundations of GR (see e.g. the historical work by Rowe (2019) and philosophical discussions in H. Brown and Brading (2002), De Haro (2021), and Freidel and Teh (2022)). This was in essence a dispute about whether theories with local symmetry could have physically meaningful conserved charges in a manner analogous to the isolated subsystems of theories with rigid symmetry.

The goal of this section is to use these two themes to provide a working definition of "substantive" – that is physically contentful, in a sense I am about to explain – general covariance, which will then be used to reconstruct the Representational RP within the setting of local symmetry in Sections 5 and 6.

"General covariance" is a huge and messy topic within the philosophy foundations of GR – I will not pretend to have scratched the surface of it (and I will not discuss topics such as "The Hole Argument" which, though interesting in their own right, are less central to my narrative path). Section 4.1 aims merely to get into our sights a minimal and uncontroversial understanding of "general covariance" for a *Lagrangian* formulation of GR – this is nothing other than what physicists colloquially speak of the diffeomorphism invariance of (the Lagrangian formulation of) GR, or – more generally – the gauge invariance of (the Lagrangian formulation of) a gauge theory. Section 4.2 then highlights how local symmetry (such as diffeomorphism or gauge symmetry) leads to two complications for the representation of focal empirical scenarios, corresponding to the following two themes: first, a complication concerning the extension of the RP; and second, a complication concerning the representation of subsystems with conserved charges which lay at the heart of the Klein-Einstein dispute. Finally, Section 4.3 uses these two complications to suggest a particular understanding of "substantive" general covariance, viz. a notion that is physically contentful enough to overcome the representational complications of Section 4.2.

4.1 A Novel Mathematical Procedure

Let us put ourselves in Einstein's shoes post SR as he contemplates the mathematical medium of symmetry *en route* to his formulation of GR. It is already clear to him how Representational RP (the representational interpretation of the Principle of Relativity that we discussed in Section 3.1) works in the context of SR and he has a rich *physics* of inertial frames in terms of which he understands RP; in other words, he knows how important the subsystem-environment conception is for understanding symmetry in physics.[25] Also clear to him is the associated mathematics of symmetry – the rigid Poincare symmetries that preserve the Minkowski metric and the form of the SR equations of motion, and which relate (mathematical) inertial coordinate-frames.

Einstein's goal is to develop a novel mathematical procedure for the physical representation of gravity, and we join him at a stage of his journey when he is fully prepared to deform the *rigid* Poincare symmetries of SR to what we now call diffeomorphism symmetry, that is smooth transformations between the (mathematical) object called a spacetime manifold and itself, and which are thus now *local* functions of the spacetime parameters. (I thus blithely skip over the prior episode of the *Entwurf* theory, when he was briefly prepared to give up on diffeomorphism symmetry!) Einstein's term for the resulting formal property was "general covariance" and he initially intended it to be an adequacy condition that would single out the theory of GR.

Einstein's physical motivations for pursuing general covariance were various and somewhat in flux. On the one hand, a key motivation came from the

---

[25] Arguably, he also understood how important the subsystem-environment conception was for correctly interpreting the Equivalence Principle, as Lehmkuhl (2023) argues.

physics of inertial frames: certain states of motion – but not others – are priv-
ileged by the inertial frame structure of SR (which, recall, are those frames
related by the Poincare group) and Einstein wanted to construct a theory in
which all states of motion were on a par – thus his attempt to "enlarge" the
Poincare group to the diffeomorphism group. On the other hand, the physics of
inertial frames (as Einstein clearly recognized in his treatment of the RP in SR)
requires an understanding of subsystem-environment relationships along the
lines of Yuyi's boat and Representational RP, and this is a point that Einstein
seems to have neglected in pursuing a different conception of general covariance, wherein a generally covariant theory is one that can be formulated in
such a way that the theory is expressible in any coordinate system (where the
different coordinate systems are related by diffeomorphisms).[26]

The problem with this latter notion of general covariance, as Kretschmann
(1918) pointedly remarked, is that it does not achieve Einstein's goal of singling
out a particular theory – just about any theory is generally covariant according to this definition, including Newtonian mechanics, as Cartan showed in his
invariant, curved geometric formulation of Newtonian gravitation (often called
Newton-Cartan theory). In light of Kretschmann's response to Einstein, it is
now commonly accepted (see Norton (2003) and references therein for discussion) that this latter notion of general covariance was a misstep. Instead, one
should try to provide a "physically contentful" or substantive notion of general
covariance that really does pick out some proper subset of physical theories,
and perhaps GR in particular. This has led to a small cottage industry of trying to articulate a notion of substantive general covariance that distinguishes it
from the mere freedom to express a theory's fields, equations of motion, and
solutions in arbitrary frames (see Pooley (2010) for a review and Freidel and
Teh (2022) for my own views on the matter).

I will return to the issue of how to distinguish substantive general covariance (or more generally: gauge symmetry) from mere expressive freedom in
Section 4.3. At present, however, I would simply like to get a minimal and
uncontroversial conception of general covariance on the table in order to fix
ideas. Before stating this version, let me provide one more stipulation about
when we are joining Einstein in his intellectual journey, namely a point at
which he is using the mathematical materials of the Lagrangian formalism in
order to formulate the theory of GR. Many readers will be familiar with this

---

[26] Actually, Blau (2011) has an interesting interpretation of general covariance on which the
notion was never meant to have any physical content in isolation, but only when conjoined with
the Einstein Equivalence principle. Interesting as the suggestion is, it – and the equivalence
principle – lie beyond the scope of our discussion.

Philosophy of Physics

formalism from a course on classical mechanics: the idea there is that instead of formulating a particle theory using the equations of motion, one introduces the integral of a functional called a Lagrangian that one tries to optimize over the space of possible particle trajectories while holding the initial and final data fixed; the optimal trajectory describes the motion of said particle (and is thus a solution to the corresponding equations of motion). In the next section, we will employ this apparatus in the case of field theory but *without* fixing the initial and final data, so that it describes the space of solutions and not just a particular solution of the theory.

With that out of the way, let us turn to stating the "minimal" version of general covariance for a Lagrangian theory:

> **(BGC)** First, distinguish between the background (non-dynamical) and the dynamical fields of the theory. We then say that a Lagrangian[27] of a theory is *basically generally covariant* (BGC) just in case diffeomorphisms of its dynamical fields are variational symmetries[28] of the Lagrangian.[29]

BGC is a notion of general covariance that is adapted to variational formulations of physical theories, and it is one that Einstein would have found uncontroversially necessary by the time he had started working on variational formulations of GR (NB: by which I do not mean that he would have found it contentful enough to count as substantive). It is also what physicists invariably mean when they speak of a "diffeomorphism invariant" Lagrangian theory.

By extension, I will also use BGC to refer to the analogous case in which diffeomorphisms are replaced by internal gauge symmetries; which case I mean will be obvious from the context. Although there are of course important differences between diffeomorphism symmetry and internal gauge symmetry, the important contrast for our purposes is between theories whose actions have only rigid symmetries (and thus do not satisfy BGC) and theories with BGC, that is whose actions have symmetries that are functions of spacetime (such as diffeomorphism symmetry and internal gauge symmetry).

---

[27] For reasons that will become clear in Sections 5 and 6 (and which are amply discussed in Freidel et al. (2020), Freidel et al. (2021)). I note that I do *not* assume that a theory has a unique Lagrangian: thus, the Einstein-Hilbert Lagrangian should be understood as just one out of a possible class of standardly stipulated Lagrangians for GR.

[28] In other words, diffeomorphisms leave the Lagrangian invariant up to exact terms; we remind the reader that, as a consequence, such diffeomorphisms also take solutions of the theory's equations of motion (EOM) to solutions.

[29] For ease of comparison, I note that BGC is the variational version of what Pooley calls "GC5" in Pooley (2010) and what he calls "Diffeomorphism Invariance (final version)" in Pooley (2017). Pooley also identifies the variational form with what he calls "Background Independence (version 2)" in Pooley (2017).

As we are about to see, the novel mathematical procedure of Einstein's general covariance, that is the local variational symmetries of BGC, led to both a complication of Einstein's search for physical representation and – ultimately – to a profound reconceptualization of his desired subject: a novel kind of empirical scenario instantiating Representational RP. In the next subsection, we turn to the complications caused by this alteration in the mathematical medium.

4.2 Complications of the Medium

There are two closely related features that Einstein took to be crucial in imparting a sense of the *subject* in the physical representations of SR (where recall, the "subject" is an empirical scenario), and which he correspondingly sought through the novel mathematical procedure of BGC. First, a conception of an isolated subsystem along the lines of Yuyi's boat or the Representational RP (which is an empirical description for the reasons that I mention in Section 1.2.1). And second, a conception of an isolated subsystem as having conserved charges, whose mathematical aspect lies in the fact that the rigid Poincare symmetries can be used to construct such mathematical quantities.

Thus, we can conceptualize part of Einstein's quest to achieve *physical* representation in GR as the struggle to understand how, through the materials of BGC, one might in some sense recover these features. The reason I speak of a "struggle" is that – as we are about to see – the materials of BGC themselves produce complications for trying to represent the empirical scenario associated with RP and the subsystem charges. At this point, I should again emphasize that in much of this tradition of representation, the environment is not represented explicitly; thus, although both the physical understanding of the RP and the notion of charge require an environment reference standard, we will not be explicitly modelling the environmental degrees of freedom in what follows. I turn now to the complications.

4.2.1 Extending the Relativity Principle

The first complication arises from the relationship between general covariance and the RP. At one point in the development of GR, Einstein thought that general covariance was simply an extension of the RP in SR. Later, as part of his response to Kretschmann's critique, he retreated to the view that general covariance is necessary but insufficient for extending RP (see Pooley (2017) for a discussion of Einstein's retreat). For our purposes, it will be more fruitful to frame Einstein's question thus: does one's preferred notion of general covariance play a distinctive role in extending the Special Relativistic

RP?[30] If "yes," then this shows us how general covariance can be used to recover the empirical subject associated with Yuyi's boat; indeed we can consider general covariance to be physically substantive by virtue of playing this role (clearly, it cannot *just* be the freedom to reexpress the representation in arbitrary coordinates if it plays this role).

Here we find Einstein in the familiar situation of making a physically suggestive conjecture that is nonetheless amenable to being problematized by others. Correspondingly, there has been much controversy about whether *any* notion of general covariance (including BGC) can play a role in extending RP. At least within the philosophical literature, one finds various authors (e.g. Belot (2000) and Norton (1993)) who answer "no" because they claim that the desired extension of the RP to GR must be trivial, in the sense that the resulting "relativity" group is just the identity. In brief, these authors argue for their claim by appealing to the following (controversial) way of thinking about RP, which I will call *Geometric RP*: in such an extension, the empirically significant group of symmetries invoked in the RP should be identified with the stabilizer group of the spacetime metric in GR (just as it is typically identified with the stabilizer group of the Minkowski metric in SR).

If one accepts Geometric RP, then since a generic metric – which is now a dynamical field – in GR has no nontrivial automorphisms, it immediately follows that the RP is trivial in GR. Thus, no form of general covariance (including BGC) can play a role in enabling such an extension. I stress that this conclusion is wholly driven by the assumption of Geometric RP: if the argument works, it works regardless of what one takes substantive general covariance to be.

On the other hand, as I discussed in Section 3.1, it is not Geometric RP but the Representational RP that captures the empirical meaning of Yuyi's boat-type scenarios. As a reminder, recall that according to H. R. Brown and Sypel (1995), the symmetry group of RP is to be understood as the symmetry group relating inertial frames, which are in turn to be fleshed out as – in idealization – ways of encoding the equivalence of outcomes of experiments set up with the same initial conditions in an *isolated* subsystem. Thus, when in SR one goes on to define a geometric object (the Minkowski spacetime metric) whose stabilizer group is precisely the symmetry group of the RP, one is merely codifying "...aspects of the comparative behaviour of different systems of physical rods and clocks in relative motion," where the behavior of the target

[30] The parallel of this question for internal symmetry is: does one's preferred conception of local internal symmetry (for a Lagrangian) play a distinctive role in extending the RP for global internal symmetry? For an illustration of the latter as "relative phase difference" in the $U(1)$ case, see Teh (2016) and Greaves and Wallace (2014).

subsystem (whose inertial frames are related by these symmetries) is being measured with respect to an environment frame, from which the subsystem is dynamically isolated. This is the essence of what I called Representational RP. (Of course, once one understands this point, one can abstract away from a concrete material system and even its idealization in terms of inertial frames, and arrive at a kind of "iconic" or "generic" image that we call "Minkowski spacetime," which provides an efficacious vehicle for abstract reasoning. This generic "type" can then be made to descend once again into a representation of a concrete empirical scenario when it is filled in with the description of the relevant subsystem-environment decomposition, initial and boundary conditions including the relevant conditions for "isolation," and the practical knowledge of the experimentalist.)

Notice that, according to Representational RP, one needs to be able to understand how a theory models an "isolated subsystem" *before* one can even articulate RP within that theory. Thus, advocates of Representational RP will not accept this argument for Geometric RP, because it is based on a definition of RP that ignores the question of how "dynamical isolation" is to be modelled in the extended scenario.[31] Furthermore, a moment's thought will show that "dynamical isolation" is a much more sophisticated notion in GR than it is in SR: we typically model it as the asymptotic flatness of a subsystem spacetime.

To recapitulate, we have arrived at the following way of sharpening Einstein's question: given an appropriate choice of boundary conditions representing an "isolated system," is there a notion of general covariance that has a distinctive role to play in extending RP – in the sense of Representational RP – from SR to GR? While Einstein was presumably here concerned with asymptotic boundary conditions for subsystems, I will note that his question can be generalized so that we consider isolated finite subsystems as well, and not just asymptotic infinity.[32]

In Section 5, we will see that Noether's theorems – when taken up into the context of physical representation – strongly suggest that the answer to the

---

[31] Ironically, Brading and Brown (2004) have expressed sympathy for Geometric RP on the grounds that the diffeomorphism symmetry of GR "...does not have an active interpretation in terms of isolated subsystems of the universe." The debate about whether and in what sense local symmetry (of which the diffeomorphism symmetry of GR is an instance) can have such an interpretation has by now played out quite fully (see Teh (2016) and references therein), but my interest here is in highlighting that, on a representational view of RP (and again: in contrast to Geometric RP), one's conception of substantive general covariance should do real work in adjudicating the matter.

[32] Furthermore (although it lies beyond the scope of this Element), it is also physically relevant to consider non-isolated or *open* subsystems in which charges are not conserved, as has been done in Freidel et al. (2021).

above question is "yes." But before that we will need to reckon with a second – closely related – facet of Einstein's struggle with general covariance, and the one that led Noether to write her 1918 paper reconceptualizing our understanding of the *mathematical* property of general covariance (and in particular of BGC).

4.2.2 The Klein-Einstein Dispute

The second complication arises when, like Einstein, we ask the question of whether the mathematical materials of general covariance can be used to represent the kind of empirical scenario – paradigmatically, an isolated subsystem! – that has physically meaningful conserved charges, such as energy. For instance, GR is a theory whose Lagrangian has the diffeomorphism symmetries of BGC: does it by virtue of this have physically meaningful charges? (We will see in Section 6 that this question is in fact of a piece with our previous question in Section 4.2.1 concerning the RP.)

Historically speaking, this question found its genesis in Hilbert's variational treatment of GR (see Rowe (2019, 2021) for a masterful treatment of the history, which I follow here). In addition to deriving the field equations in his treatment, Hilbert's main contribution (which he called "the most important goal of his theory") was the formulation of an invariant "energy vector" and a proof of its conservation. In response, Einstein provided his own slightly different variational approach to GR; in this note he derived an energy current that can be written in modern notation as $J_X = C_X + dU_X$, where X is a local symmetry, C_X is a quantity[33] that vanishes on-shell (meaning that it vanishes when the equations of motion are satisfied, that is when we are considering solutions to these equations), and U_X is called the "charge aspect" or "superpotential" (these formulae are generic for any theory with local symmetry – including GR – but we will provide explicit details for electromagnetism in the next section).

It was this series of developments that set the stage for Emmy Noether's contribution to the mathematical foundations of GR, in terms of which the complications of BGC for Einstein's representational project are best understood. While working as Felix Klein's research assistant, Noether worked out the exact relationship between Hilbert's energy vector and Einstein's energy current. This analysis, in conjunction with Klein's own work on the topic, led to an open letter from Klein to Hilbert in which Klein explained that the energy current J_X could always be decomposed as the sum of two parts, the first of which vanishes on-shell (meaning that it vanishes when the equations of motion

---

[33] More precisely, C_X is a constraint associated with the gauge transformation X.

are satisfied) and the second of whose divergence is identically zero, regardless of whether the equations of motion are satisfied (physicists usually refer to such a property as an "off-shell" property).

On the basis of this analysis, Klein asserted that the statement that $dJ_X = 0$ is merely a mathematical identity and thus does not have any physical content – hence the complication for Einstein's attempt to represent energy conservation for isolated systems using the subtle medium of BGC. Klein's point here, I take it, is that (as we will see more clearly in the next section), in the normal case of theories with a rigid symmetry, conservation laws hold on-shell only, that is $dJ \approx 0$ (where "$\approx$" denotes equality when the equations of motion are satisfied, also called "on-shell equality") and are in this sense consequences of the particular dynamics of the relevant physical theory; by contrast, in the generally covariant (or the gauge theory) case, the conservation law *seems* to hold independently of the dynamics one is trying to model and thus appears physically vacuous.

Upon reading this letter, Einstein wrote Klein to express admiration for his insights, but also to protest that "I regard what you remark about my formulation of the conservation laws as incorrect." According to Einstein, the conservation of the current was not itself a mathematical identity, but was instead the consequence of a mathematical identity *and* the equations of motion. Einstein further argued that the physical interpretation of such a conservation statement was analogous to the integral form of Gauss's law. To see what Einstein meant, it will help to have before us the cartoon of a spacetime subsystem displayed in Figure 2. Recalling that the codimension of a d-dimensional submanifold in an ambient four-dimensional spacetime is given by $(4 - d)$, we see here that the initial Cauchy surface Σ is of codimension 1, Γ is a time-like boundary of codimension 1, and $\partial\Sigma$ is a codimension 2 submanifold that is typically called a "corner".

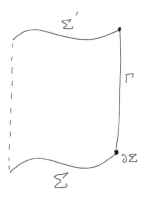

Figure 2 A subsystem with corner $\partial\Sigma$

Here is how Einstein reasoned: since the current J_X can be shown to be on-shell exact, that is $J_X \approx dU_X$ (where I remind the reader that "$\approx$" denotes equality when the equations of motion are satisfied), one can use Stokes' theorem to write the on-shell charge (the current integrated over Σ) as a *corner* quantity, namely a quantity that is only integrated over $\partial\Sigma$. In symbols, we have $\int_\Sigma dU_X = \int_{\partial\Sigma} U_X$, where Σ is a Cauchy surface and its boundary $\partial\Sigma$ is the so-called "corner" of the manifold. Einstein then proceeded to sketch a particular subsystem for whose boundary conditions (essentially a primitive version of asymptotic flatness) such a computation of charge made sense. It may help the reader here to point out the analogy between such charges and more mundane systems in which the charge of a subsystem is computed by means of quantities on its spatial boundary (where we represent the subsystem as isolated from any relevant interference), as in the use of Gauss' Law in electrostatics.

As regards Einstein and Klein, their further exchange only led to a stalemate, with Einstein insisting that his conception of energy conservation in GR was physically contentful, and Klein denying this claim on the grounds that "mathematical identities" cannot be physically contentful statements. Who was right? While Klein's general line of argumentation is tendentious at best (we frequently use mathematical identities to model various physical happenings), he has nonetheless identified a genuine disanalogy between how *dynamics* enters our understanding of conserved quantities in theories with rigid symmetries, as opposed to the role it plays in theories with gauge symmetries – a point that complicates matters for Einstein, who very much wished to maintain that analogy. We will revisit their exchange in Section 6 after we have developed the tools in Section 5 to make more sense of it (in particular, we will derive all the formulae mentioned in this subsection).

For now, what matters is that against the background of this impasse concerning physical interpretation, further progress was made with regard to the mathematical medium of GR: in 1918, Noether published her seminal paper "Invariante Variationsprobleme" Noether (1918). Herein, she proved two theorems – referred to in the physics community as Noether's first theorem and Noether's second theorem respectively – that would eventually shape how just about every physicist thinks about symmetry, and she derived a corollary (the so-called "Hilbertian assertion") that bore directly on the Einstein-Klein debate. More specifically, the second theorem powerfully reframes the notion of BGC in terms of what we would today call "Noether identities," and the corollary states that (in our terminology) Lagrangian field theories with BGC only have (in virtue of that BGC) trivial currents, where we remind the reader that a "trivial current" is one that can be written as the sum of an exact form and a term that vanishes on-shell (i.e. our C_X and dU_X, respectively). Thus, the on-shell current is identically conserved and is said to have an "improper conservation law."

It would be a mistake to understand Noether here as weighing in on the question of physical representation or interpretation; rather, one should understand her as offering a mathematician's conceptual analysis through her corollary: given that you think the concepts of BGC and trivial currents are physically relevant, you should be aware of the following mathematical relationship between them. Read this way, Noether's corollary does not answer our question about charges, but it does prompt the following specific version of it: "Does a theory with BGC have – on that basis – physically meaningful conserved charges?" where we should now understand the latter as "physically meaningful charges that stem from trivial currents," since (as Noether showed) nontrivial currents cannot be derived from the diffeomorphism (or gauge) symmetry of BGC.

4.3 The Einstein Test for "Substantive"

Let me sum up the foregoing. Introducing the mathematical medium of general covariance led to two profound representational complications for Einstein – two obstacles to what Aristotle would have called *theorein*. As a result, Einstein was embroiled in a struggle to refashion within this new image – this *theoria* – the kind of subject that he took to be a focal case of physical representation: an empirical scenario exemplifying the RP (or Yuyi's boat) and whose isolated subsystems have measurable conserved charges such as energy.

I would now like to suggest that it is fruitful to use Einstein's representational desiderata to furnish us with a test for whether a particular conception of general covariance counts as *substantive*. The loose notion of "substantive-ness" that I would like to begin with is just some notion of general covariance that is physically significant in a sense that goes beyond the mere freedom to formulate a theory in any coordinate system. Of course, there may be many ways of specifying such a notion, and I certainly do not make any pretense at an exhaustive classification. However, I take Einstein's desiderata to gesture at two ways in which a notion of general covariance can reasonably be said to go beyond the mere freedom to use arbitrary coordinates. First, if the general covariance (or gauge/diffeomorphism symmetry) in question plays an essential role in defining measurable charges for a subsystem; and second, if – given a specification of isolated boundary conditions for subsystem – the general covariance extends the RP relative to those boundary conditions.

Although I have not yet given you any reason to think that these two ways are related, they are in fact deeply interwoven, as we shall see in the next section. In anticipation of their relationship, let me introduces what I will call the "Einstein test" for a *substantive* notion of general covariance:[34]

1. The "corner charge" part of the Einstein test: does the candidate for substantive general covariance yield nontrivial corner charges?[35] Here, I include the qualifier "corner" in order to anticipate a result of the next section, namely that the charges associated to a gauge theory are necessarily corner charges.
2. The "Extending RP" part of the Einstein test: does the candidate for substantive general covariance allow us to extend the principle of relativity (in the sense of Representational RP that we discussed)?

If the answer to both questions is "yes," then the candidate notion counts as substantively generally covariant for the purposes of our investigation.

Given the status of BGC as the "received" version of general covariance (when formulated variationally), and Einstein, Klein, and Noether's own recognition of the relevance of BGC to the question of conserved charges, it is natural for us to try to run the Einstein test on BGC. To that end, the next section will explain how the raw mathematical materials of Noether's theorems can be taken up – in the art of the physicist – to develop a particularly trenchant formulation of this test. To presage our results: we will find in Section 6 that BGC does not pass the test, but also that the framework of Noether's second theorem suggests a version of substantive general covariance that does.

5 Symmetry *à la* Noether

In the last section, I used two of Einstein's representational desiderata (for local symmetry) to formulate a criterion – which I called the "Einstein test" – for whether general covariance (or local symmetry more generally) is substantive. Before saying more about whether and how this criterion is met, however, we will need to get quite a bit clearer about the representational possibilities offered by the mathematical medium of local symmetry.

Recall that when I introduced the notion of physical representation in Section 2, I stressed that we (perhaps implicitly) arrive at an understanding of the

---

[34] Of course, my point is not that the historical Einstein ever formulated such a test; it is rather that we can extrapolate such a test from some of his representational desiderata (which may even be in conflict with other aspects of his views concerning general covariance).

[35] Based on the history, one might reasonably take the corner charge part of the test to include the *conservation* of nontrivial corner charges. However, for the purposes of analytical clarity, it will be convenient for us to make the minimum criterion the existence of *nontrivial* corner charges, with conservation left as a further criterion.

Figure 3 Wedding feast of Herod

subject (the empirical scenario) in the representation both by following the procedures of the medium (the mathematics of local symmetry and Lagrangians) *as* the medium, and simultaneously grasping that the meanings and qualities of the medium apply to the subject in a *nonliteral* way. Thus, when we are in the act of receiving a representation, having our attention directed toward the medium's procedures and techniques – and thus to the gulf between the meanings of the representation's medium (*qua* medium) and the qualities of its physical subject – does not detract from our intellectual grip on the subject, but instead heightens our ability to do physics, that is to think *in* the representation.

On this view, our appreciation of the (representational) possibilities inherent in a representation turns in part on our appreciation of the techniques and procedures of its medium. As an illustration, consider the linear perspective technique, of which there was a growing awareness amongst Florentines in the lead-up to the *quattrocento*, but which did not reach its maturity till it was geometrically articulated by Brunelleschi and in this form taken up by artists such as Donatello and Masaccio. Donatello's "Wedding Feast of Herod" (see Figure 3) – a bronze relief on the baptistry of Siena's *Duomo* – is exemplary of this "taking up" of the medium's technique: here the linear perspective construction – a surface feature of the medium that properly belongs to the science of optics – is evident, but equally evident is how it is bent to a representational use; witness the energy that is delivered to the representation when – in following the "geometric pavement" construction up to its vanishing point – we are made to pass through three different scenes, which are thus united in the representation.

The situation with general covariance (or local symmetry) is somewhat similar. Einstein no doubt had a growing awareness of the properties of general covariance – conceived of as a mathematical medium – but the proper technique for handling this idea was not articulated until the contribution of Emmy Noether, the pure mathematician who first deeply understood the rhythms and textures of general covariance, and who by all accounts knew and cared rather little about the physics. Noether's mathematical understanding of general covariance was encapsulated in two theorems – often simply referred to as "Noether's theorems" by physicists – that will be bent to the ends of physical representation in this section and the next.

What were the rhythms and textures of general covariance that Noether so clearly discerned between 1916 and 1918? A full sense can only be obtained from a close reading of Noether (1918), but the gist might be conveyed as follows: although the fundamental formula of the variational calculus (equation (2)) was already well-known prior to Noether's contribution, what Noether effected with enormous fluency was a framework for using this formula to understand the implications of the invariance (or "symmetry") of a Lagrangian, and in particular the invariance of a Lagrangian under local symmetry (of which, recall, the diffeomorphisms of general covariance are in instance). Conceptually, what drove this development was Noether's novel insight into how geometrical structures (such as vector fields) on the *space of fields* – as opposed to more familiar mathematical manifolds such as "spacetime" – could be used to articulate the symmetries of a Lagrangian and well as the implications of this invariance for conserved quantities; at the computational level, the fluency of her methods was manifested in a procedure that is as powerful as it is effortless: integration by parts.

In fact, the version of Noether's theorems that is best suited to my purposes is not quite the exact thing that she proved in 1918, but rather her two theorems as they have been absorbed into a framework that makes the geometry of the space of fields very explicit – this is called the "covariant phase space" framework by physicists.[36] Although this framework is relatively cutting edge with respect to

---

[36] Mathematicians know it in a slightly different guise as the variational bicomplex, and – at a much higher level of sophistication – it is also implicit in the BV-BFV approach to field theories (see e.g. Cattaneo et al. (2014)).

the discussion of "symmetry" within the philosophy of physics, I should note that it is common coin in some parts of the GR and high energy theory literature.

I will lay out the covariant phase space framework in Section 5.1, with a special focus on the phenomenon of "corner charges" that arises for theories with local symmetry, and in Section 5.2, I will spell out the relationship between such corner charges and Representational RP (especially Greaves and Wallace's schema for Yuyi's boat-type scenarios that I discussed in Section 2).

5.1 The Covariant Phase Space

First, let me sketch the (mathematical) spacetime geometry of the subsystems that we will consider. It will suffice to consider the elementary case of a contractible four-dimensional spacetime M equipped with Minkowski metric η, and whose boundary has two kinds of pieces – a timelike boundary Γ, on the one hand, and (spacelike) initial and final data surfaces Σ and Σ', respectively, on the other (see Figure 2 for an illustration of this geometry).[37] Introducing such a timelike boundary Γ is crucial for modelling the boundary conditions of a subsystem; furthermore, one can think of such a finite boundary model as a regularized version of a subsystem with asymptotic boundary conditions.[38] I note that Σ meets Γ at a two-dimensional manifold $\partial\Sigma$, which is typically called a "corner"; mathematical objects supported on $\partial\Sigma$ are correspondingly called "corner quantities."

In this section, we will take a Lagrangian to be a spacetime four-form L; its integral on M is called the action $S = \int_M L$. Consider then the equivalence class $[L]$ of Lagrangians on M, where the equivalence relation is given by $L \sim L + d\ell$, that is two representatives of the class differ by a boundary Lagrangian ℓ (on Γ). Given one of these representatives (say L for concreteness) and fixing boundary conditions for the subsystem on Γ, we characterize the solutions (or dynamical states) of the subsystem as those field configurations such that an arbitrary variation of the action is stationary up to terms that are supported at Σ and Σ', that is

$$\delta S = \delta \int_M L = \int_\Sigma \Psi - \int_{\Sigma'} \Psi, \tag{2}$$

where we assume that the variation satisfies the boundary conditions on Γ and Ψ is a local function of dynamical and background fields on Σ and Σ'. The reason that stationarity has this form – as opposed to the more familiar $\delta S = 0$ – is

---

[37] In other cases, for example when modelling a black hole, it will also be important to consider null boundaries.

[38] On the other hand, we can start with a finite boundary model and obtain an asymptotic model through holographic renormalization, see e.g. (Chandrasekaran et al., 2021).

that we would like to use it to construct an entire space of possible solutions (corresponding to different initial data) and thus we do not place any boundary conditions on Σ and Σ'.

To see why stationary field configurations are solutions to equations of motion in the usual sense, let us first take note of the fundamental variational formula (a version of which appeared in Noether's 1918 paper)

$$\delta L = -\mathcal{E} + d\theta, \tag{3}$$

where δ should here be thought of as an exterior derivative on the space of fields, d is the usual spacetime exterior derivative, $\mathcal{E} := E_\phi \, \delta\phi$ is a field space 1-form (and spacetime 4-form) whose coefficient E_ϕ is the Euler-Lagrange quantities, and θ is a field space 1-form (and spacetime 3-form) called the pre-symplectic potential current.[39] Upon integrating (3) over M, we see that in order for (2) to be satisfied for arbitrary variations that satisfy the boundary conditions on Γ, the configuration about which we are varying needs to satisfy the equation of motion $E_\phi = 0$ (furthermore, changing L by a Γ boundary term to obtain a different representative of $[L]$ yields the same bulk equation of motion).

Matters are slightly more subtle for the $d\theta$ term in (3), which descends to Γ, Σ, and Σ' when integrated, by means of Stokes' theorem. In order to choose boundary conditions for $\theta|_\Gamma$ such that (2) is satisfied, it will help to note that θ has the generic form $A\delta B$, and thus there are two kinds of boundary conditions that we can impose in order to guarantee that $\theta|_\Gamma$ vanishes) – for instance, we can set $B|_\Gamma$ to any of some set of fixed values which kills the variation (thus yielding a family of boundary conditions) or we can simply set A to zero; in what follows we will think of boundary conditions in the former "family" sense.[40] For instance, if $\theta = A\delta\phi$, then introducing a particular Dirichlet boundary condition $\phi = \phi_0$ on Γ would amount to singling out a "leaf" in the family of Dirichlet boundary conditions on Γ. In this sense, θ is associated with a family of boundary conditions (often called a "polarization"), and since in our scheme each Lagrangian $L \in [L]$ picks out a unique θ, L is also associated with that family of boundary conditions.

---

[39] In our approach, each representative L (including a possible boundary Lagrangian piece on Γ) picks out a unique θ by means of the homotopy operators introduced in Anderson (1989); this unique θ is then the one that appears in the fundamental variational formula. I refer the reader to see De Haro (2021), Freidel et al. (2021), and references therein for the details of this construction.

[40] More generally, for a representative Lagrangian $L + dl$, we can avoid terms on Γ by setting the boundary condition to be $\theta + \delta l \overset{\Gamma}{=} dC$. Here this added generality will only be useful in Section 6.3.

The covariant (pre-)*phase space P* is the space of subsystem solutions that we have just constructed by means of a variational principle (i.e. field configurations that satisfy $E_\phi = 0$). However, it is important to notice that we will not be thinking of P as a mere set – that does not carry further structure – of subsystem solutions satisfying certain boundary conditions on Γ (as I discussed in Section 2, this is how Greaves and Wallace (2014) think of the data of a subsystem). This is because the mathematical procedure that we have just used to construct P also endows it with the rich structure of a pre-symplectic manifold, in the sense that $\delta\theta$ can be integrated over an initial data surface Σ to construct the presymplectic field space 2-form $\Omega := \int_\Sigma \delta\theta$ on the space of solutions.

At an intuitive physical level, Ω is a mathematical structure that keeps track of the distinct physical degrees of freedom of the subsystem. For instance, when a vector field (on field space) is in the kernel of Ω it can be quotiented out to eliminate the degeneracy and to arrive at a correct count of the physical degrees of freedom of the system.[41] In the absence of boundary conditions on Γ, therefore, there is no reason to expect that Ω will be independent of which data surface Σ it is defined on; it could well change from surface to surface, depending on the subsystem's interactions with the environment – such changes are usually referred to in the literature as the "leakage of symplectic flux through the boundary Γ." Nonetheless, in our setting, we have imposed "closed" boundary conditions on Γ that ensure that there is no leakage of symplectic flux, and thus that Ω is *independent* of which Σ we are integrating over. In other words, in our setting, the pre-symplectic form Ω is a property of the entire closed subsystem.

I now proceed to introduce Noether's theorems. In order to do so, let us consider the case where a vector field on field space ξ is a variational symmetry of a Lagrangian L, meaning that $L_\xi L = d\ell_\xi$, where L_ξ denotes the Lie derivative (with respect to ξ) on the space of fields.[42] The Noether current is defined as (the spacetime 3-form) $J_\xi := I_\xi\theta - \ell_\xi$, where I_ξ is the field space interior product which is here being used to contract the field space 1-form θ with the field space vector field ξ so as to yield the field space 0-form J_ξ. The Noether charge can then be constructed by integrating J_ξ over a Cauchy surface Σ, that is as $Q_\xi := \int_\Sigma J_\xi$.

---

[41] This is a somewhat naive statement: one does not have to literally quotient out in order to have a correct count of the degrees of freedom; for instance, in the BV-BFV framework of Cattaneo et al. (2014), the physical degrees of freedom are identified homologically.

[42] I will adopt a standard abuse of notation in which the notation for a gauge parameter is also used to represent the corresponding vector field on field space ξ (in which that gauge parameter appears as a co-efficient).

Suppose that our Lagrangian L has a variational symmetry ξ. We can then use Cartan's magic formula ($L_\xi = \delta I_\xi + I_\xi \delta$) and (3) to derive the formula $dJ_\xi = I_\xi \mathcal{E}$. The part of Noether's first theorem that I would like to consider here now follows immediately from noticing that if ξ is a rigid symmetry, then J_ξ is a nontrivial Noether current in the sense discussed at the beginning of this section, that is it cannot be written as the sum of an exact form and a term that vanishes on-shell. Thus, we have a part of Noether's first theorem, namely that given a rigid symmetry and its associated current J, when we go on-shell (i.e. set $E_\phi = 0$), we obtain the corresponding conservation law $dJ = 0$. This is what Klein meant when he said that (in our paraphrase), in the case of rigid symmetries, the conservation laws are a *consequence* of the dynamics of the physical theory.

On the other hand, when the symmetry ξ is a local symmetry, that is a nonconstant function of spacetime, and thus has an infinite number of parameters, we find that we can write $I_\xi \mathcal{E} = dC_\xi + \mathcal{N}_\xi$ where C_ξ is a constraint term that vanishes on-shell, and $\mathcal{N}_\xi$ is the sum of terms linear in ξ, each of which contains a differential operator acting on the Euler-Lagrange quantities (in the next section, I will carry this out explicitly for the case of electromagnetism).[43] We can then use the locality of ξ to argue that $\mathcal{N}$ must vanish as an identity – this yields the nontrivial Noether identity corresponding to the local symmetry ξ.[44] By combining this observation with $L_\xi L = dl_\xi$ and (3), we can conclude that $d(J_\xi - C_\xi) = 0$ and so the Noether current takes the form $J_\xi = C_\xi + dU_\xi$, where the "superpotential" or "charge aspect" U_ξ depends on the gauge parameter ξ in a linear but nontrivial manner. In other words, we have just demonstrated a part of Noether's second theorem, viz. that a gauge symmetry gives rise to a "trivial" Noether current, that is a current that can be written as the sum of an exact form and a term that vanishes on-shell.

There are two extremely important points to emphasize about the superpotential U_ξ at this stage:

- The superpotential U_ξ is uniquely determined by our choice of Lagrangian in the equivalence class $[L]$ and the symmetry generator ξ.[45]
- For gauge theories, that theories with BGC, the Noether charge $Q_\xi^\Sigma :=$ $\int_\Sigma J_\xi = \int_{\partial\Sigma} U_\xi$ is always a *corner* quantity (in the sense that it is being integrated over the corner $\partial\Sigma$) because the Noether current is on-shell

---

[43] For a proof of this statement, see Chapters 1 and 3 of Kosmann-Schwarzbach, Schwarzbach, and Kosmann-Schwarzbach (2011).

[44] A *nontrivial* Noether identity is one whose coefficients N^i do not vanish on-shell.

[45] See Freidel et al. (2021) for discussion and see De Haro (2021) for an explicit formula for U_ξ.

exact, and thus by Stokes' theorem, the charge is determined entirely by the superpotential U_ξ and the geometry of the corner.

To round off our discussion of Noether's theorems, let me note that at a highly schematic level, the chief moral of Noether's two theorems for Lagrangian field theory is that the space of such theories can be partitioned according to whether or not the theories possess two features, which in turn correspond to the symmetry-structure of the theories' Lagrangians. These two features are:

- Noether currents, denoted by J_ξ. A *nontrivial* Noether current is one that cannot be written as the sum of an exact form and a term that vanishes on-shell. A *trivial* Noether current is one that can be so written.
- Noether identities – denoted $\mathcal{N}_\xi$ – which are relations between the Euler-Lagrange quantities that take the form $N^i[\frac{\delta L}{\delta \phi^i}] = 0$, where the coefficient N^i is a differential operator (note that since these are mathematical identities, they hold off-shell). A *nontrivial* Noether identity is one whose coefficients N^i do not vanish on-shell. A *trivial* Noether identity is one whose coefficients vanish on-shell.

Furthermore, the correspondence between Noether currents/identities and a theory's symmetry-structure is given by Noether's two theorems (and their modern interpretations), as follows:[46]

- (Noether's first theorem) Rigid symmetries (symmetries with a finite set of parameters) are in 1-1 correspondence with nontrivial Noether currents. Furthermore, these currents are conserved when we assume the on-shell property in computing the divergence of the current.
- (Noether's second theorem) Local symmetries (symmetries with an infinite set of parameters, that is symmetries in the sense of BGC) are in 1-1 correspondence with nontrivial Noether identities. Furthermore, they yield trivial currents whose on-shell expression is conserved as a mathematical identity, that is without using the on-shell property in computing the divergence.

Thus, instead of speaking of the local symmetries – or BGC – of a physical theory, we could just as well speak of its nontrivial Noether identities; instead of speaking of the rigid symmetries of a physical theory, we could just as well

---

[46] Here my presentation of the "iff" statement is heuristic. Making this an honest theorem requires considerable care about how to define equivalence classes of symmetries and currents, as demonstrated by Olver (2000).

speak of its nontrivial Noether currents (in fact, this substitution has become standard in physics subfields such as Quantum Field Theory).

With these observations in hand, we can see that it is the "charge" – indeed "corner charge," since we are discussing gauge theories! – part of the Einstein test which is more fundamental than the "Extending RP" part of the test: if the corner charges for a gauge theory are trivial (i.e. vanishing), then there is no sense in which the RP can be extended (from a rigid precursor theory) to that gauge theory. If, on the other hand, the corner charges are nontrivial, then subject to understanding what kinds of boundary conditions model an "isolated system," it may be possible to obtain a generalized version of RP in that gauge theory.

5.2 Charges and their Relationship to the GW Schema

With the insights from Noether's theorems (as absorbed into the covariant phase space formalism) in hand, we are in a position to see why I used the term "corner charge" in formulating the first part of the Einstein test for substantive general covariance in Section 4.3: this is because the Noether charges stemming from local symmetry are always *corner* quantities, that is quantities defined on the co-dimension 2 manifold $\partial\Sigma$. This is one of the distinctive features of the mathematical medium of local symmetry that makes it so different from rigid symmetry.

We are also now in a position to understand why the "corner charge" part of the Einstein test (for substantive general covariance) and the "Extending RP" part of the test are deeply interwoven, as I claimed in Section 4.3. In order to appreciate this point in summary, let us recall a basic idea from Hamiltonian classical mechanics, where a symplectic manifold (a $2n$-dimensional smooth manifold $\mathcal{P}$ endowed with a symplectic form ω) and a Hamiltonian $H : \mathcal{P} \to \mathbb{R}$ are used to describe the time evolution of a particle. In this case, we say that H is a "Hamiltonian charge" because it satisfies the following equation:

$$-I_{X_H}\omega = \delta H, \tag{4}$$

where X_H is the vector field associated to H by ω, δ is the exterior derivative on $\mathcal{P}$, and I is a contraction on $\mathcal{P}$. Furthermore, we say that the Hamiltonian charge H generates time evolution because the integral curves of X_H describe the time evolution of the particle. Analogously, in the covariant phase space description of classical field theory, we can define the Hamiltonian generator of a subsystem symmetry as a charge that satisfies the analog of (4), and in virtue of which it is called a "Hamiltonian charge." The link, then, between the first and second parts of the Einstein test is that a nontrivial corner charge (available via general covariance if it passes the first part of the test) is a candidate for playing the role of a Hamiltonian charge that generates a subsystem symmetry that "extends RP," thus passing the second part of the Einstein test, and recovering Representational RP in theories with local symmetry.

Let us briefly see how the connection between charges and Greaves and Wallace's schema for Yuyi's boat (their version of Representational RP) plays out in a case in which the relevant charge is *not* a corner quantity, namely the case of rigid symmetry.

First, recall how the GW schema for Yuyi's boat goes: the empirically significant subsystem symmetry has to preserve the isolated boundary conditions of the subsystem, and it has to be such that the composite (subsystem and environment) state prior to and after the subsystem symmetry are *not* related by a symmetry of the composite system. And recall too an important point that I have already commented on, but wish to emphasize yet again: there is of course a mathematical gap between the GW schema and the covariant phase space framework as we have just been developing it, because the latter is not in the business of explicitly modelling the environment. Nonetheless, one is still representing an environment (albeit implicitly) in our use of this framework.[47]

With that in mind, let us now explain how Noether's charges for rigid symmetries play into this schema. For concreteness, it will help to consider an example such as the free complex scalar field Lagrangian $L = -1/2\, d\phi^* \wedge \star d\phi$ and its rigid $U(1)$ internal symmetry $\phi \mapsto \exp(-i\alpha)\phi$, where α is a constant parameter. Let $\xi = -i\alpha\, \delta/\delta\phi$ be such an infinitesimal rigid symmetry represented as a tangent vector on field space. From Noether's first theorem, the fact that ξ is a variational symmetry of L implies that we have an on-shell conserved Noether current J_ξ which can be used to define a charge $Q_\xi = \int_\Sigma J_\xi$ (albeit one that does not descend to a corner like gauge charges).

Having put in place the above as well as our understanding of an implicitly represented environment (which carries a reference standard for the phase of ϕ), it is now easy to spell out the connection to Greaves and Wallace's schema: the Noether charge Q_ξ is the Hamiltonian *generator* of the (phase space) subsystem symmetries of Greaves and Wallace's schema, in the sense that it satisfies the following analog of (4) on-shell:

$$\delta Q_\xi = -I_\xi \Omega, \tag{5}$$

[47] For a discussion of how to explicitly represent the environment in the case of gauge theories on manifolds with a finite boundary, see Gomes (2021).

and the flow of its associated vector field implements the symmetry as a canonical transformation on phase space. In other words the Hamiltonian charge Q_ξ generates the relative phase difference between subsystem and environment that is discussed in Greaves and Wallace (2014). A similar point could be made for other Lagrangian theories with rigid symmetries, and to which Greaves and Wallace's schema for Yuyi's boat applies.

What then of the (variational) local or gauge symmetries of Lagrangian theories? Do these also invariably yield nontrivial *corner* charges that act as the Hamiltonian generator of the relevant subsystem symmetries? Supposing that local symmetry here yields a nontrivial corner charge (a point that we shall investigate on further in the next section) this matter is somewhat subtle: as Chandrasekaran et al. (2021) and Freidel et al. (2021) argue, in the absence of any boundary conditions, one does not obtain (5) but instead the charge-flux relation:

$$\delta Q_\xi - \mathcal{F}_\xi = -I_\xi \Omega, \tag{6}$$

where the symplectic flux term $\mathcal{F}_\xi$ measures the degree to which the corner charge Q_ξ fails to be Hamiltonian. In our setting, we have imposed boundary conditions that kill off this symplectic flux term, thus making the charge Hamiltonian relative to that boundary condition. Thus, assuming that we have chosen the right "isolated" boundary conditions for Representational RP, our discussion provides a general strategy for reconstructing RP in the case of generally covariant theories.

It is worth pausing to reflect on our earlier characterization of physical symmetry as a coarse-grained description of the empirical dynamics of a subsystem vis-a-vis a possibly implicit environment. Greaves and Wallace's schema – which represents dynamics under the aspect of the equations of motion – is one expression of this characterization, and we have just seen that the notion of variational symmetries – in which context dynamics is represented under the aspect of a Lagrangian – offers yet another expression; the two are linked by a sufficiently rich representational understanding of what it means for a subsystem charge to generate a subsystem symmetry.

5.3 Further Reading

For an introduction to the covariant phase space from a physicist's perspective, I recommend Donnelly and Freidel (2016), Freidel et al. (2020, 2021), and Harlow and Wu (2020). A somewhat more careful treatment of analytical issues is given in Khavkine (2014).

In Section 5.2, we have been assuming the imposition of non-gauge-invariant boundary conditions; for a discussion of gauge-invariant boundary conditions, see Carrozza and Hoehn (2021); Mathieu, Murray, Schenkel, and Teh (2019); Mathieu and Teh (2021); Gomes and Riello (2017) were the first to give a general discussion of abstract gauge-invariant boundary conditions within the physics context. The most general treatment of "virtual" or Segal boundaries for gluing has been given in Cattaneo et al. (2014), and a particular case (in homological degree zero) is discussed in Gomes, Hopfmüller, and Riello (2019).

6 The Subject Resolved

In the previous section, we worked through the covariant phase space (and Noether's) technique for handling the subtle mathematical medium of local symmetries (i.e. the formalism of the theory has BGC, in either the internal gauge symmetry sense, or the diffeomorphism sense). And by developing a feel for this technique, we came to grasp the momentum that it imparts to the representation – Einstein's variational formulation of GR. In particular, Noether's theorems impressed upon us the distinct and novel possibility of taking "corner charges" up into our representational purposes. Of depiction, Podro writes that the "...recognition of the subject is extended and elaborated by the way its conditions of representation, the medium and the psychological adjustments the painting invites become absorbed into its content" (Podro (1998): 2). I would like to suggest that we should understand the work of the previous section and this one in a similar vein: the conditions of representation made salient by Noether's theorems (and Noether corner charges, the possibility of their being Hamiltonian generators, etc.) and the corresponding adjustments they invite of us – *qua* physicists – to use them in pursuing the empirical subject of RP are themselves absorbed into the content of the representation.

In this section, we will continue to bring into view what Einstein so earnestly desired to bring into view but only managed to glimpse through a mirror dimly: the reconstruction of a certain physical subject – a Yuyi's boat-type empirical scenario – through the medium of local symmetry.

As background to summarizing my argument, let us recall the working definition of *substantive* general covariance that I gave by means of the "Einstein test" in Section 4.3, and fill it in a bit more in terms of the Noetherian machinery introduced in the last section. This definition of substantive general covariance had two parts. A local symmetry (of a Lagrangian) is substantively generally covariant just in case (i) (the Corner Charge part) it yields a nontrivial Noether (corner) charge by means of Noether's second theorem; and (ii) (the Extending RP part) given an appropriate choice of "isolated" boundary conditions for the subsystem, this nontrivial Noether charge is also a Hamiltonian charge that generates a subsystem symmetry for Representational RP, or – in a different idiom – a subsystem symmetry that is empirically significant in the sense of Yuyi's boat. Since I will use this notion of substantive general covariance in my arguments, it will help to explicitly state why we are justified in working with this notion: this is because one of Einstein's desiderata for general covariance was that it would extend the RP to the case of local symmetry, and we have now seen in Section 5.2 why the notion of nontrivial corner charges is essential – from the Noetherian perspective – to constructing the subsystem symmetries of any potential RP for theories with local symmetry. If the corner charge were trivial, then it would not generate any subsystem symmetries at all, and the corresponding local symmetry could not be said to – even potentially – extend RP.

I now proceed to the arguments of this section. I will first argue that BGC is not sufficient to achieve substantive general covariance. This follows from two premises:

P1 The "corner charge" part of the definition of substantive general covariance, which requires nontrivial corner Noether charges.

P2 There are two examples of theories that have been engineered to satisfy BGC but which only have trivial corner Noether charges.

In fact, the two examples in P2 are focal cases of the result of "Kretschmannizing" theories with rigid symmetries, that is modifying them so that they now have local symmetry – in the sense of BGC – but intuitively still have the same physical content as the theories premodification. Thus, a corollary of this argument is that the focal cases of Kretschmannized theories do not satisfy substantive general covariance in our sense.

Next, I will show that focal examples of "standard" theories with local symmetry do satisfy substantive general covariance in our sense. This follows from two premises:

K1 The ("corner charge" and "Extending RP" parts of the) definition of substantive general covariance.

K2 The focal examples of "standard theories with local symmetry" have nontrivial corner Noether charges, and we can impose appropriate "isolated" subsystem boundary conditions such that these Noether charges are also Hamiltonian generators for the subsystem symmetry that is relevant to RP.

I will regiment our discussion as follows. Since the case of (the diffeomorphism invariance of) GR is more technically involved, I will start by giving an explicit discussion of some elementary models in Section 6.1.

First, in Section 6.1.1, I discuss the case of a Kretschmannized scalar field theory on Minkowski space that satisfies the formal definition of BGC (for diffeomorphism symmetry), but which turns out to yield a trivial superpotential and thus fails the "corner charge" part of the Einstein test. I also sketch how essentially the same phenomenon occurs in a case where our Kretschmannization procedure is used to engineer not local diffeomorphism symmetry of the theory, but local internal $U(1)$ symmetry. Thus, this discussion establishes premise P2.

Second, in Section 6.1.2, I discuss the paradigmatic case of an elementary theory with BGC (in contrast to the physically marginal Kretschmannized cases of Section 6.1.1), namely electromagnetism. As we will see, this case yields a nontrivial superpotential and thus passes the "corner charge" part of the Einstein test for substantive general covariance. Section 6.2 then discusses the obvious suggestion for supplementing BGC in order to rule out the Kretschmann-style cases, and further discusses how the corner charge of electromagnetism can be used to pass the "extending RP" part of the Einstein test for substantive general covariance. Section 6.3 then draws on the morals of previous sections to provide a qualitative discussion of the case of GR, which is not only more technically challenging, but also contains significant conceptual subtleties as regards extending the RP. Taken together, these results establish the premise K2.

Finally, in Section 6.4 I sketch how one might draw on more sophisticated developments of Noether's technique, that is on innovations in handling the representation's mathematical medium, to provide a more thoroughgoing response to Kretschmann.

6.1 Test Cases

Let us now run the "corner charge" part of the Einstein test on BGC, that is pose the question: does a theory with BGC thereby have nontrivial corner charges? To that end, it will be helpful to use as our test cases some of the toy models that Pooley (2010, 2017) and others use to test the validity of different definitions of general covariance, although I warn the reader that for reasons of simplicity, I have opted for a slightly different labeling convention from that of Pooley (2017).

Consider first the following theory of a free scalar field on Minkowski space that we will call[48] **SR1**[49]:

$$L(\phi, \eta) = -\frac{1}{2}\eta^{ab}\partial_a\phi\partial_b\phi, \tag{7}$$

where the scalar field ϕ is the dynamical variable and the Minkowski metric η is a nondynamical (or background) structure. Our variational formula (3) for this theory is $\delta L = E_\phi\delta\phi + \nabla_a\theta^a$, where the Euler-Lagrange expression is $E_\phi = \Box\phi$ and the pre-symplectic potential current is $\theta^a = -\nabla^a\phi\delta\phi$. Notice that the conservation of the stress-energy tensor T^{ab} follows from applying Noether's first theorem to the rigid spacetime symmetries of this theory.

Clearly, this theory does not satisfy BGC, because diffeomorphisms are not a variational symmetry of ϕ. In our next example, I will modify **SR1** slightly to show that one can produce an example of a theory that satisfies BGC, but which fails the "corner charge" part of the Einstein test, because it only has a trivial superpotential, and thus the corner charge vanishes.

6.1.1 BGC with Trivial Superpotential

Consider now a theory that we will call **SR2** which is given by,

$$L^{\text{SR2}}(\phi, Y; \eta) = L(\phi, Y^*\eta), \tag{8}$$

where $Y : \mathbb{R}^d \to M$ is a parametrization field that captures our freedom to choose coordinates on the spacetime M.[50] In particular, we can use Y to express this special relativistic theory in non-inertial frames if we wish. Due to this interpretation of Y, it is appropriate to call **SR2** a "Kretschmannized" version of **SR1**. Our dynamical variables in this theory will be ϕ and Y, whereas the Minkowski metric η still retains its status as a non-dynamical background structure.

We can again compute the variational formula (3) for this theory, thus obtaining $\delta L^{\text{SR2}} = E_\phi\delta\phi + E_a\chi^a + \nabla_a\theta^a$, where $\chi := \delta Y \circ Y^{-1}$, $E_a = -2\nabla_b T^{ba}$, and $\theta^a = 2T^{ab}\chi_b - \partial^a\phi\delta\phi$. Notice that in addition to the previous equation of motion $E_\phi = 0$, we now have a new equation of motion $E_a = 0$, that is the conservation equation for the stress-energy tensor. Furthermore, by inspecting θ, we can see that in addition to the original symplectic pair ϕ and its momentum $\nabla^a\phi$, we now have a new symplectic pair Y^a and its momentum T^{ab}.

---

[48] Here I have switched to tensorial (instead of differential forms) notation in order to more conveniently express the stress-energy tensor.
[49] Pooley (2017) calls the non-variational formulation **SR1**.
[50] Pooley (2017) calls this **SR5**, and his example of **SR4** is similar.

We now draw the reader's attention to an interesting point: clearly, L^{SR2} is generally covariant in the sense of BGC (this is true even though η remains a non-dynamical field and is not varied) and thus this theory must have a nontrivial Noether identity. By turning the crank of the Noether machine, we find that this identity is $E_\phi \partial_a \phi - 2\nabla_b T^{ba} = 0$ (that this is an identity can be straightforwardly checked from the definition of T_{ab}). In other words, the content of the Noether identity **SR2** is actually the content of Noether's first theorem for **SR1** – one might call this the *transmutation* of Noether's first theorem into Noether's second theorem via Kretschmannization!

It is also interesting to compute the (trivial) Noether current, i.e. $J_\xi^a = I_\xi \theta - \iota_\xi L^{SR2} = 0$. In other words, the current vanishes identically off-shell, and we can deduce that the superpotential U_ξ – and thus the corner charge algebra – is trivial. Hence, the theory **SR2** demonstrates why BGC fails the "corner charge" part of the Einstein test for substantive general covariance (and in consequence fails the "extending RP" part of the test): even though this theory clearly has BGC, it nonetheless fails to deliver nontrivial corner charges.

The example is exemplary, and it is easy to construct analogous examples of Kretschmannized theories with a similar structure. For instance, consider the case of a scalar field theory coupled to a flat background $U(1)$ gauge field, but now equipped with a parametrization field that captures our freedom to change our typical representation of this background gauge field from $A = 0$ by switching to a "non-inertial" choice of section (which one can think of as an "internal frame" if one wishes). The Lagrangian corresponding to this case is $L(\phi, \theta) = -1/2\, d_{d\theta}\phi \wedge \star d_{d\theta}\phi$, where θ is a "parameterization field" that acts on a gauge field A as $\theta^* A = A - d\theta$ (thus shifting our representation from the usual choice $A = 0$ to $A = -d\theta$, for instance) and which transforms under the $U(1)$ gauge transformation $A \mapsto A + d\chi$ as $\theta \mapsto \theta + \chi$; and where $d_{d\theta}$ denotes the gauge covariant derivative. (Note that in this case, the "inertial frames" are precisely the ones where θ is constant – thus stabilizing $A = 0$ – and one simply has the ordinary exterior derivative, whereas the non-inertial frames are the ones where θ is nonconstant and so one is forced to introduce the covariant derivative.) Upon performing the covariant phase space analysis of this example, one finds that it replicates the structure of the previous example, in the sense that it clearly satisfies BGC (for $U(1)$ gauge symmetry) but only yields a trivial superpotential and thus also a trivial corner charge.

To sum up, this section has established P2, namely that there are two examples of theories which satisfy BGC (because they are Kretschmannized versions of theories that do not satisfy BGC), but which nonetheless only have trivial corner charges.

6.1.2 BGC with Nontrivial Superpotential

Next, we turn to a familiar example in which BGC does yield a nontrivial corner charge (and thus passes the "corner charge" part of the Einstein test). Consider the Maxwell gauge theory Lagrangian

$$L = -\frac{1}{4}F \wedge \star F, \tag{9}$$

where we have chosen a representative $L \in [L]$ with vanishing boundary Lagrangian, that is $\ell = 0$. The variational formula (3) is

$$\delta L = -E\,\delta A + d\theta, \tag{10}$$

where δ is the exterior derivative on the space of fields, $E = d \star F$ is the Euler-Lagrange expression, and $\theta = \star F\,\delta A$ is the presymplectic potential current.

In this case, the gauge symmetry vector field on field space has the form $\hat{X} = dX(x,t)\delta/\delta A$, and $L_{\hat{X}}$ is the field space Lie derivative with respect to this vector field, although we will now abuse notation by using X to denote both the field space vector field and the local gauge parameter. Since $L_X L = 0$, we can repeat the general analysis summarized in Section 3.1 to obtain

$$0 = L_X L = d(I_X \theta) - I_X(E\delta A) \tag{11}$$

and then by using integration by parts and the definition of the current $J_X = I_X\theta$, we have

$$d(J_X - EX) = -X\,dE. \tag{12}$$

Since X is an arbitrary gauge symmetry (i.e. a local function of spacetime), we can integrate both sides over a domain with boundary, apply Stokes' theorem to the LHS, and – by assuming that X and its derivatives vanish on the boundary – deduce that $dE = 0$ as an identity. This is a simple example of a nontrivial Noether identity (here the Noether operator N is just the exterior derivative d, which does not vanish on-shell). We note that $dE = dd \ast F = 0$ is clearly a mathematical identity since it follows from $d^2 = 0$.

By applying the Noether identity $dE = 0$ and the Poincare lemma, we see from (12) that

$$J_X = EX + dU_X \approx dU_X, \tag{13}$$

where "$\approx$" denotes "on-shell equality."[51] In other words, we see that the current J_X that is associated with the gauge symmetry is trivial, although – as we

---

[51] This is nothing other than the $U(1)$ gauge theory version of Einstein's formula $J = \mathfrak{T} + \mathfrak{t}$, of which the reader can find an analysis in De Haro (2021). It explains why energy pseudotensors in GR are often said to be defined only up to a superpotential.

are about to see – this of course does not imply that the Noether charge is trivial!

Given that we have a current J, we can define a charge by integrating J over a spacelike Cauchy surface Σ to obtain:

$$Q^{\Sigma} = \int_{\Sigma} J. \tag{14}$$

We now remind the reader that a unique possibility arises for *trivial* currents (but not nontrivial currents), which is a direct consequence of the *nontrivial* Noether identity: by Stokes' theorem, a trivial on-shell current $J_X \approx dU$ can be converted into a purely boundary Noether charge

$$Q_X^{\Sigma} \approx \int_{\Sigma} dU_X = \int_{\partial\Sigma} U_X = \int_{\partial\Sigma} X(\star F), \tag{15}$$

where in the last equality we used the explicit form of the superpotential U_X (fixed by X and our choice of L), which – unlike the case of **SR2** – is nontrivial and leads to a nontrivial corner charge algebra.

To go further and say more about the conservation of this corner charge and its role as Hamiltonian generator, one needs to explicitly spell out the subsystem (canonical) boundary condition that we have been assuming in our analysis, namely that the pullback of θ to Γ vanishes. Recall that such a boundary condition corresponds to our particular choice of L, and also rules out the leakage of "symplectic flux" from the subsystem (see Freidel et al. (2021); Harlow and Wu (2020) for a more comprehensive discussion of flux leakage).[52] In the context of the present example, the pre-symplectic potential current is $\theta = \star F \delta A$, and so the corresponding boundary condition is $\delta A|_{\Gamma} = 0$.[53] This is a Dirichlet boundary condition for the gauge field A, whose standard interpretation is precisely as representing an *isolated* subsystem, for which the electric charge should be conserved.[54]

The symmetries of the subsystem need to preserve this boundary condition, and so we should set $dX|_{\partial M} = 0$. Evidently, when $X|_{\partial M} = 0$, the symmetry is in the kernel of Ω and so does not represent a physical symmetry generated by a charge, leaving non-zero constant transformations at the boundary

---

[52] In general, one also needs to make sure that the chosen boundary configuration has stabilizers, although this condition is guaranteed in our example, since all $U(1)$ gauge field configurations have stabilizers (the constant gauge transformations).

[53] This is of course a gauge non-invariant boundary condition; the story of how to lift it to a gauge-invariant boundary condition is somewhat more involved, so I will only touch on it briefly in the Epilogue.

[54] Of course, there are other boundary conditions one could consider, for example Dirichlet or mixed boundary conditions, that would be appropriate for modelling other kinds of subsystems. For a discussion of the suitability of fixing the electric and magnetic flux on the boundary (instead of the gauge field value), see Freidel et al. (2021).

X as the symmetries generated by the corner charge Q_X^Σ (notice that these are precisely the nontrivial stabilizers of the boundary condition).[55] Thus, we find that in the present example, the Noether corner charge is also the Hamiltonian charge associated with our boundary condition of interest – the Dirichlet boundary condition – and it generates "rigid on the boundary" $U(1)$ symmetry transformations in phase space. We have thus produced an example in which we are able to connect the nontrivial corner charges of a gauge subsystem to the subsystem symmetry transformations of Greaves and Wallace's schema in the manner discussed in Section 5.2 – indeed, such boundary rigid subsystem symmetries are precisely what they appeal to when they argue that gauge theories can instantiate Yuyi's boat-type scenarios. However, as we will soon see, there can be more complex scenarios in which a Hamiltonian charge associated with a particular boundary condition is the Noether charge of a shifted Lagrangian $L + d\ell$, but not of the original Lagrangian L that we considered. In Section 6.3, I will briefly discuss how one can resolve this tension between Noether and Hamiltonian charges – which is relevant to the "extending RP" part of the Einstein test for substantive general covariance – but let us first collect results from the present discussion.

6.2 The RP and Charges

We have just seen that, due to examples such as **SR2**, theories with BGC might still fail the "corner charge" part of the Einstein test for substantive general covariance. On the other hand, we have also seen that central examples of theories with BGC (electromagnetism, and by a simple extension: Yang–Mills theory and GR) do yield nontrivial corner charges. The general moral is that while BGC is insufficient for the existence of nontrivial corner charges, it is nonetheless necessary, because one needs a trivial Noether current in order to obtain a *corner* charge. What then needs to be added to BGC to obtain a nontrivial corner charge?

As it turns out, this is one of the cases where the analysis of the difficulty (by means of the Noether machine) also makes the remedy obvious: what needs to be added is precisely the requirement that the theory's Lagrangian L and the variational symmetry ξ be such that the resulting superpotential is nontrivial (I again remind the reader that a choice of L and its variational symmetry ξ also fixes the superpotential U_ξ). In Freidel and Teh (2022), Laurent Freidel and I thus proposed the following definition of substantive general covariance in response to the "corner charge" part of the Einstein test:

---

[55] For a much more comprehensive discussion of stabilizers at a boundary, see Gomes et al. (2019).

(Corner-SGC) A theory has Corner-SGC when, in addition to BGC, its Lagrangian L and variational symmetry ξ yield a nontrivial superpotential U_ξ.[56]

It is worth pausing to reflect on how Noether's second theorem has brought us to this insight. On the one hand, the potency of Noether's second theorem here is not that of a purely mathematical result. However, just as the character of a line or a stroke can begin to suggest the animation or dynamism of a figure in painting, so too can mathematics be taken up into the mobilization of physical thought. What Noether's second theorem does here is to offer physicists a heightened awareness of what the mathematical medium of field theory (understood through the lens of the variational calculus and Noether's theorems) is capable of when it is pressed into the service of physical representation: our discussion thus far is precisely an illustration of how the second theorem can be so recruited! Through that discussion, we come to see that the substantiveness of general covariance is not just about the local symmetry of a Lagrangian, but also about how – through that symmetry – some kinds of Lagrangians determine a nontrivial corner charge algebra.

I now turn to the "Extending RP" part of the Einstein test, for which (Corner-SGC) is a necessary but insufficient condition, because of the aforementioned potential tension between the Noether charge and the Hamiltonian charge relative to some boundary condition. Note that for a gauge theory (where I mean to include both GR and Yang–Mills theory), whatever the symmetries of generalized Representational RP are, they will need to be generated by the nontrivial corner charges from (Corner-SGC). However, there is an added complication present in generalizing RP: we are now considering the symmetries of dynamically isolated subsystems, which are generated by the Hamiltonian charges for the boundary conditions that model dynamical isolation – thus, we will only succeed in extending RP if the corner charges coming from the trivial Noether current of some Lagrangian are also Hamiltonian charges relative to these isolated boundary conditions. In Section 6.1.2, we saw an example (Maxwell gauge theory with Dirichlet boundary conditions) where this coincidence clearly obtains: we can think of it as a regularized (finite) version of a case in which the asymptotically rigid $U(1)$ symmetries of some subsystem provide a generalized version of RP relative to an environment subsystem (where the environment has been idealized away into the boundary conditions). However, as we are about to see in the next section, the corner charges from some

---

[56] Furthermore, the corresponding Noether charge will be conserved if we impose boundary conditions – consistent with the Lagrangian – that rule out flux leakage.

particular Lagrangian can also fail to be Hamiltonian charges relative to our desired choice of dynamically isolated boundary conditions. I thus propose the following more demanding notion of substantive general covariance in order to satisfy the "extending RP" part of the Einstein test:

> **(RP-SGC)** Select boundary conditions that are supposed to represent "dynamical isolation" in the context of some theory. The theory's Lagrangian L has RP-SGC just in case, in addition to satisfying Corner-SGC, the Noether charge coming from L is also a Hamiltonian charge relative to the "dynamically isolated" boundary conditions.

In the next subsection, we will consider the case of GR, which illustrates the subtleties of RP-SGC.

6.3 Revisiting the Case of General Relativity

In Freidel et al. (2021), the general way in which Noether charges can fail to be Hamiltonian charges is considered, and this framework is then applied to the case of GR (Section 3.4 of Freidel et al. (2021)). Here I only summarize the relevant features for us to revisit the Einstein-Klein dispute, namely the second of the "complications of the medium" that I mentioned in Section 3.

First, consider the Einstein-Hilbert Lagrangian L_{EH}. We can turn the crank of the covariant phase space recipe sketched in Section 5 and compute the (nontrivial) superpotential corresponding to L_{EH}, which yields the Noether (corner) charge that is known as the Komar charge (see (2.38) of Freidel et al. (2020) for a definition). This shows that we have satisfied (Corner-SGC).

Next, let us consider whether this Lagrangian satisfies (RP-SGC). To do so, suppose that we wish to use Dirichlet boundary conditions to model an isolated system (or a finite boundary regularization thereof): that is to say, we wish to set the induced metric $\bar{g}_{ab}$ on the timelike boundary Γ to a fixed value, thereby also setting $\delta\bar{g}_{ab} \overset{\Gamma}{=} 0$. More specifically, and as per standard practice, we will take $\bar{g}_{ab}$ to be the Minkowski metric, thus leading to an "asymptotically flat" solution in the limit where we take the boundary to infinity.[57] It is well-known that given this Dirichlet boundary condition, one can construct a Hamiltonian charge; however, this is not the Komar charge, but a different quasi-local charge known as the Brown-York charge (see (2.31) of Freidel et al. (2020) for a definition).

---

[57] I stress that one reason to choose the Minkowski metric as the boundary condition is that it has nontrivial stabilizers, and so one can construct subsystem symmetries that are not in the kernel of Ω and still preserve the boundary condition; a similar point could be made for (non-Abelian) Yang–Mills theory. Both cases are unlike (Abelian) Maxwell theory in the sense that a generic field configuration will not have stabilizers.

Thus, we have just encountered the complication that we discussed earlier: if one chooses L_{EH} as one's Lagrangian and takes the Dirichlet boundary condition as the "isolated boundary condition" in the "Extending RP" part of the Einstein test, then one ends up with a mismatch between the Noether charge (the Komar charge) corresponding to L_{EH} and the Hamiltonian charge (the Brown-York charge) corresponding to the Dirichlet boundary condition. The resolution of this complication, which is discussed in Freidel et al. (2021); Harlow and Wu (2020), is to choose a different Lagrangian – which is associated with a different boundary condition – in order to satisfy (RP-SGC). In particular, if we choose the Lagrangian $L = L_{EH} + d\ell_{GHY}$, where ℓ_{GHY} is the celebrated Gibbons-Hawking-York boundary term, then the corresponding Noether charge is precisely the Brown-York charge, so we have now brought our Lagrangian into alignment with the boundary condition – and the Hamiltonian charge – that we are after. Furthermore, as Harlow and Wu (2020) shows, if we assume that the metric tends toward the Minkowski metric in the asymptotic limit, then we recover precisely the ADM charges for an asymptotically flat spacetime, which is precisely what one expects of an isolated subsystem in GR. In particular, these charges generate the "asymptotic Poincare symmetries" at spatial infinity, which are the generalized analogs of the rigid Poincare symmetries (from the standard RP) in this context.[58] We thus see that $L = L_{EH} + d\ell_{GHY}$ satisfies (RP-SGC), but L_{EH} does not. In other words, despite the subtlety associated with making a careful choice of Lagrangian here – and which (RP-SGC) is designed to take into account – we can conclude that the subject of Yuyi's boat can indeed be reconstructed within the setting of GR, and that it once again connects up with the Greaves and Wallace schema in the manner discussed in Section 5.2 (namely, the Brown-York charge is the Hamiltonian generator of subsystem symmetries of the isolated subsystem).

With these clarifications in hand, let me close this subsection by returning to an assessment of the aforementioned dispute between Einstein and Klein (I refer the reader to De Haro (2021) for a complementary analysis of this dispute, with an emphasis on holographic renormalization). To what extent was Einstein confused about the status of conserved charges in GR, and to what extent did he have the representational insights into whose service we have just pressed Noether's second theorem? To answer this question, let us consider several key points from Einstein's March 1918 response to Klein (letter 480 in Einstein (1998)):

---

[58] Here we are only discussing spatial infinity; once one takes into account null infinity, one is forced to consider the infinite-dimensional BMS group if one wants a rich set of solutions.

- Einstein says the importance of the quantity $J_\xi = \mathfrak{T} + \mathfrak{t}$ (the on-shell exact Noether current) is its physical interpretation as the energy of a point mass when we go very far away from the subsystem; in other words, when we treat the subsystem as dynamically isolated from an environment.
- He also says that the possibility of this physical interpretation is underwritten by the fact that J_ξ (on-shell) is an *exact* term, presumably because he knows that via Stokes' theorem, a bulk integral of an exact term can be transformed into a boundary integral (which is why we call such charges "quasi-local").
- Einstein emphasizes that the equations of motion are necessary in order to show that J_ξ is exact, and thus J_ξ should be regarded as physically meaningful.

And then – a little over a week later – in letter 492 to Klein, Einstein (1998) goes on to place a heuristic "isolated" boundary condition on the subsystem that allows for the construction of a conserved charge!

We can now ask ourselves how Einstein's more physically oriented (and more mathematically imprecise) strategy looks with the clarity of hindsight. It would seem that Einstein already understood much of the intuitive physical content that can be expressed by means of Noether's second theorem: certainly, he understood that the *distinctive* qualitative feature of a gauge theory (implied by Noether's second theorem) is the on-shell exactness of the Noether current, and that this feature leads to a boundary charge; and he was undoubtedly sensitive to the need to *choose* boundary conditions in order to construct such a charge. In all this, I believe that Einstein was vindicated in resisting Klein's criticisms.

On the other hand, the place where Einstein's understanding was most lacking was in his grasp of (i) the relationship between a choice of a Lagrangian L and a choice of boundary conditions, (ii) the potential mismatch between the Noether charge corresponding to some L and the Hamiltonian charge corresponding to some choice of boundary conditions, and (iii) the potential appearance of infinite-dimensional symmetries (such as the BMS group[59] at null infinity) in the asymptotic limit. Indeed it is perhaps this last feature and – more generally – the physics of null infinity for an asymptotically flat subsystem that inclines me to say that the search for Yuyi's boat in GR does not so much result in a replication of the subject, but a novel and radical reworking of it. Yes, we discern very clearly the contours of Yuyi's boat at the spatial

---

[59] The BMS group is an enhanced group of asymptotic symmetries (over and above the familiar asymptotic Lorentz symmetries) at null infinity. Mathematically speaking, it is the semidirect product of the Lorentz group with an infinite-dimensional Abelian group.

boundary of the subsystem (as we have just seen above), but at the same time, asymptotically flat subsystems bring with them the physics of gravitational radiation, and as Wieland (2021) has recently shown, the corresponding interpretation of null infinity as an *open* Hamiltonian system.[60] And yet, such novelty is precisely what one should expect from the impulse that the procedures of mathematics can give to the task of physical representation: for the reworking of a subject like Yuyi's boat is not something essentially static – a replication as it were – but a dynamic mobilization of the physical imagination in search of new understanding, and of new phenomena to understand.

6.4 A Response to Kretschmann

In this last subsection, I should like to return to Kretschmann's critique of a conception of "general covariance" as the ability to express a physical theory in an arbitrary frame, whether we are talking about a spacetime coordinate frame (as in the case of a special relativistic scalar field theory) or an internal gauge frame (as in the case of a scalar field coupled to a flat gauge potential). Recall that Kretschmann's objection is that, on this conception, there is nothing about general covariance that would distinguish the ones we would like to call *gauge* theories (or diffeomorphism-invariant theories), because all physical theories (in which a notion of frame applies) can be expressed in an arbitrary frame.

For instance, in the case of a scalar field theory formulated on a background Minkowski space, we can certainly choose to formulate its equations of motion in a non-inertial frame obtained via pulling back the metric and fields by a general diffeomorphism – the resulting equations would not have the simple form that we are used to from working in inertial frames (for they would pick up connection coefficients), but nothing would have changed concerning the physics that such a formalism is trying to represent. And as I mentioned at the end of Section 6.1.1, we can make a similar point in the case of a scalar field theory formulated relative to a background flat $U(1)$ gauge field A: there is an analog of "inertial" internal frames, viz. those gauge frames in which the exact (here I assume a contractible spacetime) $U(1)$ gauge field vanishes, and these are of course all connected by *global* $U(1)$ transformations which are the stabilizers of the flat background gauge field, just as the Poincare transformations are the stabilizers of the Minkowski metric.[61] But here again, and similarly to

---

[60] In other words, the condition of asymptotic flatness does not after all guarantee a truly "isolated" system as one might naively guess – there can still be energy exchange with the environment at null infinity.

[61] Recall that a "stabilizer" of a mathematical object is a transformation that leaves the object unchanged.

the spacetime case, one has the freedom to re-formulate the equation of motion $d \star d\phi = 0$ in a non-inertial frame obtained via transforming $A = 0$ by an arbitrary $U(1)$ gauge transformation. The new equations would not take the simple form that we are used to (they would now in general involve a gauge covariant derivative and a non-vanishing exact background gauge field) but the physics would not have changed for all that.

Kretschmann is right so far as this goes. But equipped with the materials of the previous subsections, we can now make a deeper response.

The response begins by emphasizing that a substantive form of general covariance is *not* after all the ability to express a physical theory in an arbitrary frame – this is not how Einstein used the diffeomorphism symmetry of GR in practice, and it is not how future generations of physicists would (again in practice!) use the gauge symmetry of Yang–Mills theory and other gauge theories. One route toward this point stems from the results of the previous subsections: xx′ consider a minimal description of general covariance/gauge invariance/diffeomorphism invariance such as BGC, and note that one can satisfy this description by simply rewriting a theory with rigid symmetries using parametrization fields (as we saw in Section 6.1.1) and then allowing oneself to vary the parametrization fields and to take into account their local gauge symmetries. But as we saw earlier, this kind of repackaging does not have the capacity to yield the nontrivial corner charges that would allow us to (given appropriate boundary conditions implementing subsystem isolation) generate the empirically significant symmetries of a Yuyi's boat type scenario – plausibly, any such Kretschmannizing will yield trivial corner charges. Thus, we can conclude that for the practicing physicist, substantive general covariance requires that, in addition to BGC, the theory have at least the representational capacity for the construction of nontrivial corner charges.

Further to this point, is it possible to say anything more incisive about what – formally speaking – really distinguishes theories with "substantive" gauge symmetry from the mere freedom to rewrite the theory in an arbitrary frame? For instance, what kind of *formal* standard of comparison might one use to articulate this point? While the full details of an account lie beyond the scope of this Element, it is possible to sketch a strategy for answering these questions, based on the work of Cattaneo et al. (2014) and Mathieu et al. (2019).

Recall that in the previous Subsections, Noether's second theorem (as it is conceptualized within the covariant phase space formalism) gave us control over the more subtle distinctions that we needed in order to distinguish Kretschmannized theories with local symmetry from theories with physically substantive forms of local symmetry – the latter marked by their non-vanishing corner charges. Noether's theorems spring from the manipulation (and geometrical interpretation) of the fundamental variational formula $\delta L = -\mathcal{E} + d\theta$; in order to go further in pursuing our strategy, we will need to embark on a radical reimagining of the concept of a variation itself and what it means for a variation to vanish.

We are already familiar with the idea of finding the space of solutions in the standard variational calculus (without boundaries): one sets $\delta S = \delta \int_M L = 0$ and thereby goes "on-shell." The radical reimagining in question is motivated by the insight that this naive way of making δS "coincide" with zero is unduly sensitive to mathematically different – but physically equivalent – reformulations of the space of fields, in the sense that these reformulations will in general produce mathematically distinct spaces of solutions (at least according to naively chosen mathematical standards of distinctness), unaccompanied by the (mathematical) expressive resources to describe their physical equivalence. For instance, in the case of the Kretschmannized theories that we have discussed, setting $\delta S = 0$ clearly produces a distinct space of solutions from their un-Kretschmannized versions. The resolution of this formal difficulty is that one can try to construct the "space" of solutions by means of derived intersection theory, in which case this novel "space" – called the derived critical locus – will have the mathematical form of a "stack" within which one really does have the mathematical resources to express the kind of physical equivalence that we have been discussing.[62] In particular, this framework has the formal resources to express the observables of a theory – and thus the Noether charges – in a "homological" way that is invariant under superficial mathematical reformulations. The strategy, then, for applying this framework to our scenario would be to introduce a precise enough definition of Krestchmannization to prove a result along the following lines: any Krestchmannized theory gives rise to the same "solution stack" as its pre-Krestchmannized version, and in particular, these theories share the same observables.

6.5 Further Reading

The examination of our test cases originated in a context in which substantive general covariance was being discussed in relation to the notion of "background independence." For the early discussion of these cases, see Pooley (2010, 2017) and references therein, and see Freidel and Teh (2022) for a reassessment of the

---

[62] Such "stacks" remember information about how symmetry transformations identify various solutions without the need to take a naive quotient, which loses such information – information that is essential for gluing – and produces nasty singularities.

relationship between general covariance and background independence in light of the results of this section.

7 Epilogue: The Road Forward

Although the discussion of this Element has been restricted to cases in which we impose gauge-non-invariant boundary conditions (such as setting the gauge field A equal to a fixed 1-form on the boundary), it is possible to generalize this to gauge-invariant boundary conditions (such as letting A be gauge equivalent to a fixed 1-form the boundary); and although we have (for good reason!) spent much time discussing Yuyi's boat-type empirical scenarios, there are also other cases in which physical symmetry plays a crucial role, such as that of Newton's Corollary VI. These generalizations and extensions are bound up with present research at the frontiers of physical symmetry, and it would be a shame not to give the reader some sense of how our story can be elaborated to include these aspects. To that end, I leave you with a four-part sketch of how these elements fit together.

7.1 Donnelly-Freidel Edge Modes

First, the reader might well wonder how the covariant phase space framework might need to be modified if – hypothetically – one did not place any (gauge-non-invariant) boundary conditions on the fields. This is precisely the investigation that Donnelly and Freidel (2016) undertook in the case of Yang–Mills theory and GR. What they found is that if one allows arbitrary field-independent gauge transformations at the boundary Γ of a spacetime M, then one ends up with a "symplectic anomaly" – the pre-symplectic form Ω is no longer invariant under gauge transformations. To restore gauge-invariance, they extended the phase space by introduced new "edge mode" degrees of freedom (and their momenta), as well as a corresponding corner symplectic form. They also argued that a new kind of physical "boundary symmetry" should manifest itself at the corner (call this the DF-type boundary symmetry).

Donnelly and Freidel's extension of the phase space was based on requirement of restoring gauge-invariance, but they did not specify the boundary Lagrangian that led to these edge mode degrees of freedom. Subsequently, in Mathieu et al. (2019) (see also Geiller and Jai-Akson (2020)), we showed that kinematically, these edge modes come from imposing a topological boundary condition (i.e. a fixed trival boundary bundle) on the subsystem, and that the extended symplectic structure is produced by specifying a particular boundary (edge mode) Lagrangian on Γ. Furthermore, Mathieu and Teh (2021) argued that this boundary Lagrangian could be understood as a kind of spontaneous symmetry breaking on the boundary, and that the DF-type symmetries could be interpreted as the result of acting on the boundary condition by means of an "external automorphism," in the sense that the transformation is carried out by an observer in the environment of the subsystem (or whose observational scale includes the environment).

7.2 Classical Spontaneous Symmetry Breaking

The next part of the story involves the notion of what Strocchi (2011) calls *classical* spontaneous symmetry breaking (SSB). To start with, consider the conceptual elements of what we would generally regard as an SSB scenario in condensed matter physics: there is a degenerate set of ground states – often taken to define "superselection sectors" because they cannot be changed by any subsystem operation – all connected by the "broken" symmetry transformations; there is an "internal" (or subsystem) observational scale which is unable to discriminate between which of these ground states the subsystem observer is in (although the subsystem observer can detect small fluctuations around the ground state); and there is an "external" (or environmental) scale with respect to which these ground states are in fact distinguishable. A key objective in applying the SSB formalism to this kind of scenario is to produce an *effective field theory* of the small fluctuations about a ground state; indeed this can be done very generally using the "coset construction," which allows us to construct an effective Lagrangian using only the symmetry-breaking pattern (and without detailed knowledge of the underlying dynamics, such as one has to hand in the classic example of Ginzburg-Landau theory).

This framework can be carried over quite straightforwardly to the setting of classical field theory, with the modification that here one should think of a particular subsystem boundary condition of a certain type (e.g. Dirichlet) as defining a superselection sector, in the sense that the subsystem observer is not (let us arrange) able to perform any operations that can change the boundary condition. Then, if the relevant boundary conditions are related by a subgroup of the symmetry group of the theory, we can apply the standard SSB apparatus in this context. Thus, for instance, we can the SSB coset construction to construct a classical free point-particle Lagrangian as the effective dynamics of a scenario where boosts and translations are spontaneously broken. This background will be very useful to understand the next part of the story.

7.3 Edge Modes from Reference Frames

More recently, Carrozza and Hoehn (2021) have posed the question of how one might understand the *operational* meaning of the boundary edge modes (and extended phase space) introduced in Donnelly and Freidel (2016). The following scenario – which can be generalized quite greatly – is exemplary of the kind of answer that they give.

Let M be a subsystem spacetime with timelike boundary Γ, which divides it from an environment spacetime $\bar{M}$ (strictly speaking, the environment spacetime is the complement of M in an ambient spacetime). One of the examples that Carrozza and Hoehn consider is a $U(1)$ gauge theory set on the total spacetime $M \cup \bar{M}$, where $\bar{M}$ is assumed to have an asymptotic boundary where gauge transformations fall off sufficiently rapidly – this boundary provides a non-dynamical anchor for Wilson lines and can be used to represent for example the heavy degrees of freedom of a measurement apparatus. They start with a dynamical environment gauge field $\bar{A}$ on $\bar{M}$ and observe that by taking a Wilson line of $\bar{A}$ from a reference point on the asymptotic boundary to a point $x \in \Gamma$, one thereby constructs a $U(1)$-valued reference frame on Γ, which is precisely what Donnelly and Freidel call an edge mode. This edge mode on Γ, then, is a way of "internalizing" to Γ the dynamical reference frame degrees of freedom provided by environment, so that even if one proceeds to omit $\bar{M}$ from the formalism of the representation, one can still has a formal way of expressing how the field degrees of freedom on M relate to the field degrees of freedom on $\bar{M}$ – in particular, the edge mode helps us to encode the relational information (between subsystem and environment) that allows us to reconstruct all the gauge-invariant observables with support on $M \cup \bar{M}$ (including ones that we would not have been able to reconstruct from just the individual regional data).

In the $U(1)$ case, it is more convenient to work on the Lie algebra level, in which case the edge mode can be expressed as $\varphi \in \Omega^0(\Gamma)$ (the group-valued reference frame being $U = \exp(i\varphi)$) and transforms as $\varphi \mapsto \varphi + \chi$ under a boundary gauge transformation $A \mapsto A + d\chi$. As discussed in Donnelly and Freidel (2016); Mathieu et al. (2019) one can use this edge mode field to dress the boundary gauge field, thus obtaining a (boundary) gauge-invariant field $a := A|_\Gamma - d\varphi$ (which was already introduced in Mathieu et al. (2019) in order to write the boundary symmetry-breaking action). As Carrozza and Hoehn (2021) observe, this dressed field a is a relational observable, describing the value of $A_{|\Gamma}$ (a subsystem degree of freedom) when the $U(1)$ dynamical reference frame (defined using environment degrees of freedom) is in some orientation.

Carrozza and Hoehn go on to describe the process of imposing gauge-invariant boundary conditions, that is boundary conditions for the dressed field a, on the subsystem. They do this for Dirichlet, Neumann, and mixed boundary conditions, but here I will simply illustrate the Dirichlet case: take the space of total solutions $\mathcal{S}$ over $M \cup \bar{M}$ and foliate it into a set of leaves, each of which is defined by the gauge-invariant Dirichlet boundary conditions $a = X_0$ (where

X_0 is some fixed background 1-form field), so that $\mathcal{S} = \bigsqcup_{X_0} \mathcal{S}_{X_0}$, where a leaf $\mathcal{S}_{X_0}$ is the set of field configurations in $\mathcal{S}$ such that $a = X_0$. They then note that we can proceed to focus just on the subsystem dynamics over M, defined by a particular boundary condition $a = X_0$: "Since the [boundary condition defines] the dynamical theory for M [...] we can think of each leaf in the foliation of $\mathcal{S}$ as a particular subregion theory. As such, the global solution space $\mathcal{S}$ assumes the role of a space of subregion theories, in this sense constituting a meta-theory for the local subregions."

Indeed, once we follow Carrozza and Hoehn (2021) in the last step, we see that there are two kinds of transformations that are supported on the subsystem boundary Γ. First, we have the gauge-transformations $A \mapsto A + d\chi, \varphi \mapsto \varphi + \chi$, $a \mapsto a$, which lie in the kernel of the extended pre-symplectic form constructed in Donnelly and Freidel (2016) and Mathieu et al. (2019) – these do not encode any relational difference between the subsystem and the environment because one is transforming the gauge field and the edge mode simultaneously. Second, we have the reference frame re-orientations $A \mapsto A, \varphi \mapsto \varphi - \rho, a \mapsto a + d\rho$, which can be further divided into two types. When ρ stabilizes a (i.e. is a constant, in this case), it preserve the gauge-invariant Dirichlet boundary condition, and we have what I earlier called a DF-type boundary symmetry – this symmetry *does* explicitly capture a relational difference between the subsystem and environment degrees of freedom (where the latter are encoded in the edge mode). On the other hand, a generic spacetime-dependent ρ will not stabilize a: it will instead take us from the leaf whose boundary condition is $a = X_0$ into a different leaf in whose boundary condition is $a = X_0 + d\rho$. Carrozza and Hoehn call these "meta-symmetries" because they do not transform a solution to a solution within a subregion theory, but instead transform one subregion theory into a different subregion theory (in Carrozza and Hoehn (2021)'s use of "theory").

But now this picture should look somewhat familiar based on our previous sketch of Classical SSB: as Teh and Mathieu (forthcoming) have recently observed, in this case, Carrozza and Hoehn have essentially defined a Classical SSB scenario for $U(1)$ gauge theory, where the subsystem leaves define a degenerate set of superselection sectors of the scenario that are related by non-constant local $U(1)$ gauge transformations, and each superselection sector is stabilized by a global $U(1)$ subgroup. In other words, the symmetry-breaking pattern is that of local $U(1)$ to global $U(1)$. As a sanity check, we can go in the reverse direction: we can start with just this symmetry-breaking pattern and easily compute the effective dynamics within a superselection sector by means of the coset construction. Reassuringly, at lowest order, the effective Lagrangian is precisely that of a free $U(1)$ gauge theory – precisely the dynamics from which (with the gauge-invariant Dirichlet boundary condition) we defined this symmetry-breaking pattern to begin with!

7.4 Yuyi's Boat and Corollary VI

What does any of this have to do with Yuyi's boat-type scenarios and Newton's Corollary VI? To see the connection with Yuyi's boat, I refer the reader to toy mechanical model for edge modes introduced in Carrozza and Hoehn (2021), where they use a particle degree of freedom in the environment to define an edge mode reference frame, and use another particle as a stand-in for the "boundary." For gauge-invariant Dirichlet boundary conditions on the subsystem particle degree of freedom $x^i(t)$, they show that any uniform (Galilean) boost of the particle relative to the edge mode will take one boundary condition to another, and so in the classical SSB terminology that we have been developing, this is a scenario where the Galilean boost symmetry is completely broken (since there are no nontrivial stabilizers). Nonetheless, an internal observer (Yuyi running short-range experiments on this boat) is still unable to distinguish between different superselection sectors.

To see the connection with Newton's Corollary VI, consider that the (pure) potential part of the Newtonian gravitational Lagrangian

$$ S = \int_{\mathbb{R} \times M} \left(\frac{1}{2} m \dot{\mathbf{u}}^2 - m\phi\left(\mathbf{r},t\right) \right) \delta^{(3)}\left(\mathbf{u}\left(t\right) - \mathbf{r}\right) - \frac{1}{8\pi G}\left(\boldsymbol{\nabla}\phi\right)^2 d^3\mathbf{r}dt \qquad (16) $$

is very reminiscent of the $U(1)$ gauge theory Lagrangian, except that in the former the exterior derivative and Hodge star are purely spatial, and the resulting equations of motion are elliptic (and fixed completely by the boundary data). By running an analysis that is analogous to the $U(1)$ gauge theory analysis that we just sketched, one can define an edge mode (using the frame of a body in the environment) for a Newtonian gravitational subsystem and show that gauge-invariant Dirichlet boundary conditions for the potential lead to a Classical SSB scenario from the *local* in time boosts of Newton's Corollary VI to the constant (or global) Galilean boosts of Yuyi's boat. And, in fact, when we think hard about it, what else is Newton's Corollary VI but a statement of SSB in this sense? For consider again what Newton writes: "If bodies are moved in any way among themselves, and are urged by equal accelerative forces along parallel lines, they will all continue to move among themselves in the same way as if they were not acted on by those forces." Newton of course knew that in speaking of this physical indistinguishability, he was describing it from the perspective of a subsystem observer equipped with a certain measurement scale (think of local measurements made in the vicinity of Jupiter and its moons) and that there could also be an "external" observational scale according to which different uniform accelerations could be discriminated (say, observations of the Jupiter subsystem relative to the sun). The indistinguishability that Newton is discussing is thus the inability of the subsystem observer to discriminate between a degenerate set of SSB superselection sectors, which are related to each other by (nonconstant) time-dependent boosts, and stabilized by Galilean boosts. And again very reassuringly, the coset construction tells us that the symmetry-breaking pattern from time-local to time-global boosts results precisely in the dynamics of the Poisson equation at lowest order. Thus, we see that our present line of thought has managed to unify both Yuyi's boat and Newton's Corollary VI under the rubric of edge modes and Classical SSB.

References

Anderson, I. M. (1989). *The variational bicomplex* (Tech. Rep.). Utah State Technical Report, 1989, http://math.usu.edu/~fg~mp.

Anscombe, G. E. M. (1971). *Causality and determination: An inaugural lecture*. CUP Archive.

Belot, G. (2000). Geometry and motion. *British Journal for the Philosophy of Science*, *51*(4), 561–595.

Belot, G. (2018). Fifty million Elvis fans can't be wrong. *Noûs*, *52*(4), 946–981.

Blau, M. (2011). *Lecture notes on general relativity*. Albert Einstein Center for Fundamental Physics Bern.

Brading, K., & Brown, H. R. (2004). Are gauge symmetry transformations observable? *British Journal for the Philosophy of Science*, *55*(4), 645–665.

Brown, H. (2005). *Physical relativity: Space-time structure from a dynamical perspective*. Clarendon Press. https://books.google.com/books?id=LbAUDAAAQBAJ

Brown, H., & Brading, K. (2002). General covariance from the perspective of Noether's theorems. *Diálogos*, 59–86.

Brown, H. R., & Sypel, R. (1995). On the meaning of the relativity principle and other symmetries. *International Studies in the Philosophy of Science*, *9*(3), 235–253.

Callender, C., & Cohen, J. (2006). There is no special problem about scientific representation. *Theoria. Revista de teoría, historia y fundamentos de la ciencia*, *21*(1), 67–85.

Carrozza, S., & Hoehn, P. A. (2021). Edge modes as reference frames and boundary actions from post-selection. *arXiv preprint arXiv:2109.06184*.

Cartwright, N. (1999). *The dappled world: A study of the boundaries of science*. Cambridge University Press. https://books.google.com/books?id=tOFv_i9oiAgC

Cattaneo, A. S., Mnev, P., & Reshetikhin, N. (2014). Classical bv theories on manifolds with boundary. *Communications in Mathematical Physics*, *332*(2), 535–603. https://dx.doi.org/10.1007/s00220-014-2145-3

Chandrasekaran, V., Flanagan, E. E., Shehzad, I., & Speranza, A. J. (2021). A general framework for gravitational charges and holographic renormalization. *arXiv preprint arXiv:2111.11974*.

Chang, H. (2022). *Realism for realistic people*. Cambridge University Press.

De Haro, S. (2021). Noether's theorems and energy in general relativity. *arXiv preprint arXiv:2103.17160*.

Delacrétaz, L. V., Endlich, S., Monin, A., Penco, R., & Riva, F. (2014). (re-) inventing the relativistic wheel: Gravity, cosets, and spinning objects. *Journal of High Energy Physics, 2014*(11), 1–31.

Donnelly, W., & Freidel, L. (2016). Local subsystems in gauge theory and gravity. *Journal of High Energy Physics, 2016*(9). https://dx.doi.org/10.1007/JHEP09(2016)102.

Einstein, A. (1905). On the Electrodynamics of Moving Bodies. *Annalen der Physik*, 17, 891–921.

Einstein, A. (1998). *Collected papers of Albert Einstein: The Berlin years (trans suppl)* (Schulmann, Ed.). Princeton University Press.

Freidel, L., Geiller, M., & Pranzetti, D. (2020). Edge modes of gravity. Part i. corner potentials and charges. *Journal of High Energy Physics, 2020*(11), 1–52.

Freidel, L., Oliveri, R., Pranzetti, D., & Speziale, S. (2021). Extended corner symmetry, charge bracket and Einstein's equations. *Journal of High Energy Physics, 2021*(9), 1–38.

Freidel, L., & Teh, N. (2022). Substantive general covariance and the Einstein -Klein dispute: A Noetherian approach. In J. Read & N. Teh (eds.), *The philosophy and physics of Noether's theorems.* Cambridge University Press.

Frigg, R., & Hartmann, S. (2020). Models in science. In E. N. Zalta (ed.), *The Stanford encyclopedia of philosophy* (Spring 2020 ed.). Metaphysics Research Lab, Stanford University. https://plato.stanford.edu/archives/spr2020/entries/models-science/

Frigg, R., & Nguyen, J. (2020). *Modelling nature: An opinionated introduction to scientific representation.* Springer.

Galilei, G. (1967). Dialogue concerning the two world systems. *Drake,(trans.), Berkeley CA: University of California Press.(Original work published in 1632).*

Geiller, M., & Jai-Akson, P. (2020). Extended actions, dynamics of edge modes, and entanglement entropy. *Journal of High Energy Physics, 2020*(9), 1–57.

Gomes, H. (2019). Gauging the boundary in field-space. *Studies in History and Philosophy of Science Part B: Studies in History and Philosophy of Modern Physics*, 67, 89–110.

Gomes, H. (2021). Holism as the empirical significance of symmetries. *European Journal for Philosophy of Science, 11*(3), 1–41.

Gomes, H., Hopfmüller, F., & Riello, A. (2019). A unified geometric framework for boundary charges and dressings: Non-abelian theory and matter. *Nuclear Physics B, 941*, 249–315. https://dx.doi.org/10.1016/j.nuclphysb.2019.02.020

Gomes, H., & Riello, A. (2017). The observer's ghost: Notes on a field space connection. *Journal of High Energy Physics, 2017*(5), 1–31.

Greaves, H., & Wallace, D. (2014). Empirical consequences of symmetries. *The British Journal for the Philosophy of Science, 65*(1), 59–89.

Harlow, D., & Wu, J.- q. (2020). Covariant phase space with boundaries. *Journal of High Energy Physics, 2020*(10), 1–52.

Khavkine, I. (2014). Covariant phase space, constraints, gauge and the Peierls formula. *International Journal of Modern Physics A, 29*(05), 1430009.

Kosmann-Schwarzbach, Y., Schwarzbach, B. E., & Kosmann-Schwarzbach, Y. (2011). *The Noether theorems.* Springer.

Kretschmann, E. (1918). Über den physikalischen sinn der relativitätspostulate, a. einsteins neue und seine ursprüngliche relativitätstheorie. *Annalen der Physik, 358*(16), 575–614.

Lehmkuhl, D. (2023). *Einstein's principles: On the interpretation of gravity.* Oxford University Press.

Martens, N. C., & Read, J. (2020). Sophistry about symmetries? *Synthese.* https://philsci-archive.pitt.edu/17184/

Martz, L. (1990). *Thomas more: The search for the inner man.* Yale University Press. https://books.google.com/books?id=R45XD3gZIt0C

Mathieu, P., Murray, L., Schenkel, A., & Teh, N. J. (2019). Homological perspective on edge modes in linear Yang–Mills and Chern–Simons theory. *arXiv preprint arXiv:1907.10651.*

Mathieu, P., & Teh, N. (2021, Jul). Boundary electromagnetic duality from homological edge modes. *Journal of High Energy Physics, 2021*(7). https://dx.doi.org/10.1007/JHEP07(2021)192 doi: 10.1007/jhep07(2021)192

McCraw, D. (1986). *The poetry of Chen Yuyi.* Stanford University PhD dissertation.

Nguyen, J., Teh, N. J., & Wells, L. (2020). Why surplus structure is not superfluous. *The British Journal for the Philosophy of Science, 71*(2), 665–695.

Noether, E. (1918). Invariante variations probleme, math-phys. *Klasse, 1918,* 235–257.

Norton, J. D. (1993). General covariance and the foundations of general relativity: Eight decades of dispute. *Reports on progress in physics, 56*(7), 791–858.

Norton, J. D. (2003). General covariance, gauge theories and the Kretschmann objection. In K. Brading & E. Castellani (eds.), *Symmetries in physics: Philosophical reflections* (pp. 110–123). Cambridge University Press.

Olver, P. J. (2000). *Applications of lie groups to differential equations* (Vol. 107). Springer Science & Business Media.

Podro, M. (1987). Depiction and the golden calf. In A. Harrison (ed.), *Philosophy and the visual arts* (pp. 3–28). Royal Institute of Philosophy Conferences, vol. 4. Springer.

Podro, M. (1998). *Depiction*. Yale University Press. https://books.google.com/books?id=QKaf2f30lrMC

Pooley, O. (2010). Substantive general covariance: Another decade of dispute. In M. Suárez, M. Dorato, & M. Rédei (eds.), *EPSA philosophical issues in the sciences* (pp. 197–209). Springer.

Pooley, O. (2017). Background independence, diffeomorphism invariance and the meaning of coordinates. In D. Lehmkuhl, G. Schiemann, & E. Scholz (eds.), *Towards a theory of spacetime theories* (pp. 105–143). Birkhauser.

Potochnik, A. (2020). *Idealization and the aims of science*. University of Chicago Press. https://books.google.com/books?id=CL4lEAAAQBAJ

Ramírez, S. M., & Teh, N. (2020). Abandoning Galileo's ship: The quest for non-relational empirical significance. *The British Journal for the Philosophy of Science*. https://philsci-archive.pitt.edu/17429/

Read, J. (2018). Explanation, geometry, and conspiracy in relativity theory. https://philsci-archive.pitt.edu/15253/ (Submitted to C. Beisbart, T. Sauer and C. Wuthrich (eds.), "Thinking about Space and Time: 100 Years of Applying and Interpreting General Relativity", Einstein Studies Series, Basel: Birkhauser, 2019.).

Rovelli, C. (2014). Why Gauge? *Foundations of Physics*, *44*, 91–104.

Rowe, D. E. (2019). Emmy Noether on energy conservation in general relativity. *arXiv preprint arXiv:1912.03269*.

Rowe, D. E. (2021). *Emmy Noether–mathematician extraordinaire*. Springer.

Scruton, R. (1997). *The aesthetics of music*. Oxford University Press.

Stevens, S. (2020). Regularity relationalism and the constructivist project. *The British Journal for the Philosophy of Science*, *71*(1), 353–372.

Strocchi, F. (2011). Spontaneous symmetry breaking in classical systems. *Scholarpedia*, *6*(10), 11195.

Teh, N. J. (2015). A note on Rovelli's "why gauge?". *European Journal for Philosophy of Science*, *5*, 339–348.

Teh, N. J. (2016). Galileo's gauge: Understanding the empirical significance of gauge symmetry. *Philosophy of Science*, *83*(1), 93–118.

Van Fraassen, B. C. (2010). *Scientific representation: Paradoxes of perspective*. Oxford University Press.

Vasari, G. (1900). *The lives of the painters, sculptors & architects* (No. v. 4). J. M. Dent. https://books.google.com/books?id=ZUQX5tbgAKgC

Wallace, D. (2019). *Observability, redundancy and modality for dynamical symmetry transformations.* https://philsci-archive.pitt.edu/18813/ (Revised 3/2021 to correct a few typos and add a section on Noether's Theorem.).

Wallace, D. (2021a). Isolated systems and their symmetries, part i: General framework and particle-mechanics examples. https://philsci-archive.pitt .edu/19728/ (Revised version; some typos, technical errors, and stylistic infelicities corrected.).

Wallace, D. (2021b). *Isolated systems and their symmetries, part ii: Local and global symmetries of field theories.* https://philsci-archive.pitt.edu/19729/ (Revised version: Corrects various typos, technical errors, and stylistic infelicities.).

Wieland, W. (2021). Null infinity as an open Hamiltonian system. *Journal of High Energy Physics*, *2021*, 95. https://doi.org/10.1007/JHEP04(2021)095.

Acknowledgments

The writing of this Element would not have been possible without the love, support, and counsel of many, to whom my gratitude is due. To my wife Madeleine and our children Samuel, Dorothea, Josephine, and Esther: the writing of a book may well be leisure (as Pieper tells us), but you make this leisure worth having. To my parents Heng Ong and Geraldine, who taught me the importance of tradition and hard work: thank you for your support and love through all these years.

Next, I should like to record a sincere debt of gratitude to various teachers, colleagues and collaborators who have been fellow travelers and interlocutors in this intellectual journey: Feraz Azhar, Harvey Brown, Jeremy Butterfield, David Cory, Therese Cory, Laurent Freidel, Don Howard, Niels Linnemann, John O'Callaghan, Oliver Pooley, James Read, Ira Rothstein and David Wallace. I thank David Baker for a very careful reading of this Element and numerous helpful suggestions, as well as an anonymous reader for feedback.

Finally, I gratefully acknowledge the support of JTF grant 61521 and NSF grant 1947155 while working on this Element.

"For my parents."

Cambridge Elements ⁼

The Philosophy of Physics

James Owen Weatherall
University of California, Irvine

James Owen Weatherall is Professor of Logic and Philosophy of Science at the University of California, Irvine. He is the author, with Cailin O'Connor, of *The Misinformation Age: How False Beliefs Spread* (Yale, 2019), which was selected as a *New York Times* Editors' Choice and Recommended Reading by *Scientific American*. His previous books were *Void: The Strange Physics of Nothing* (Yale, 2016) and the *New York Times* bestseller *The Physics of Wall Street: A Brief History of Predicting the Unpredictable* (Houghton Mifflin Harcourt, 2013). He has published approximately fifty peer-reviewed research articles in journals in leading physics and philosophy of science journals and has delivered over 100 invited academic talks and public lectures.

About the Series

This Cambridge Elements series provides concise and structured introductions to all the central topics in the philosophy of physics. The Elements in the series are written by distinguished senior scholars and bright junior scholars with relevant expertise, producing balanced, comprehensive coverage of multiple perspectives in the philosophy of physics.

Cambridge Elements $\equiv$

The Philosophy of Physics

www.ingramcontent.com/pod-product-compliance
Ingram Content Group UK Ltd.
Pitfield, Milton Keynes, MK11 3LW, UK
UKHW020405180125
453697UK00007B/143